COLLISION WITH HISTORY

COLLISION WITH HISTORY
The Search for John F. Kennedy's
PT 109

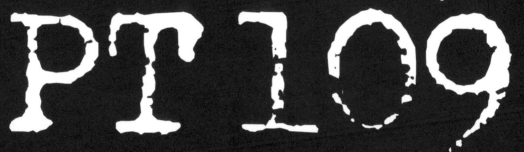

ROBERT D. BALLARD
with Michael Hamilton Morgan

INTRODUCTION BY SENATOR EDWARD M. KENNEDY

NATIONAL GEOGRAPHIC

Washington, D.C.

Skipper Lt. (jg.) John F. Kennedy (far right) and some of his crew stand easy on the deck of
PT 109 in the Solomon Islands, western Pacific, in July 1943. In less than a month, fast and
fragile *PT 109* was to be rammed, split, and sunk by a Japanese destroyer in a night action. Back
row: L-R, Al Webb (a friend not in the crew), Leon Drawdy, Edgar Mauer, Edmund Drewitch,
John Maguire, Kennedy. Front row: L-R, Charles Harris, Maurice Kowal, Andrew J. Kirksey,
Leonard Thom.

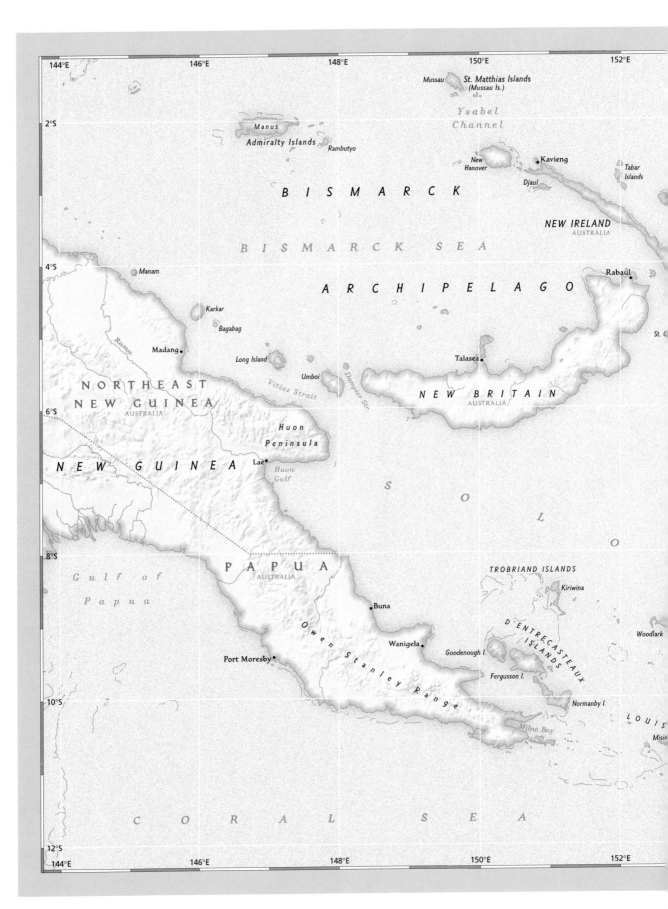

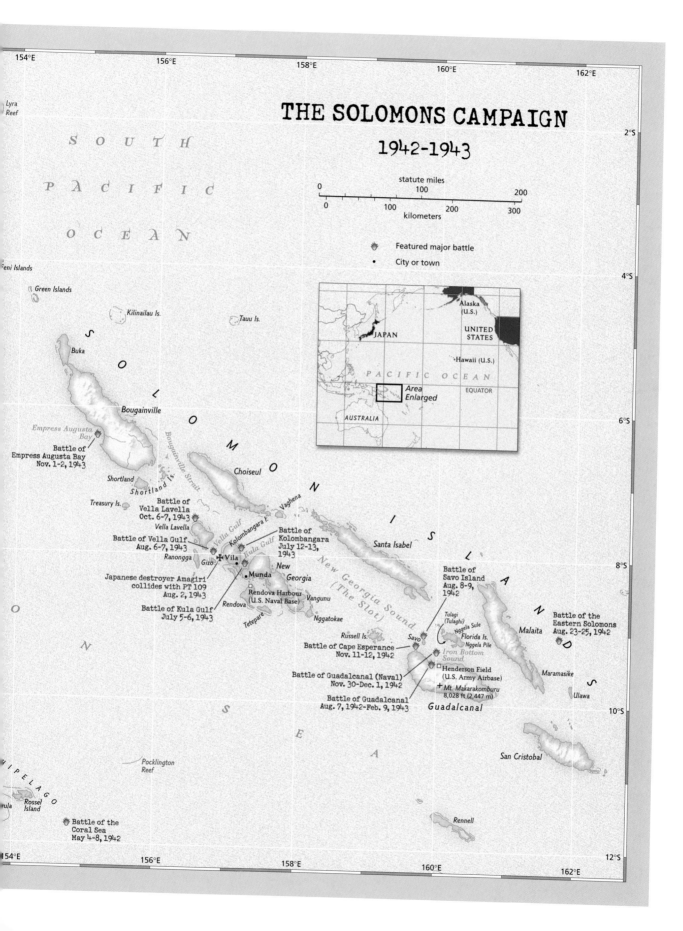

THE SOLOMONS CAMPAIGN
1942-1943

154°E 156°E 158°E 160°E 162°E

2°S

Lyra Reef

S O U T H

P A C I F I C

O C E A N

statute miles
0   100   200
0   100   200   300
kilometers

Feni Islands

4°S

🌺 Featured major battle
• City or town

Green Islands

Kilinailau Is.

Tauu Is.

Alaska (U.S.)

JAPAN

UNITED STATES

Hawaii (U.S.)

S

Buka

O

L

P A C I F I C   O C E A N

Area Enlarged

EQUATOR

AUSTRALIA

6°S

O

Bougainville

M

Empress Augusta Bay
Battle of Empress Augusta Bay
Nov. 1-2, 1943

Bougainville Strait

O

Choiseul

N

Santa Isabel

I

Shortland
Shortland Is.

Treasury Is.

Battle of Vella Lavella
Oct. 6-7, 1943 🌺
Vella Lavella

Vaghena

S

8°S

Battle of Vella Gulf
Aug. 6-7, 1943

Vella Gulf
Kolombangara I.
Kula Gulf

Battle of Kolombangara
July 12-13, 1943

Ranongga
Gizo ✠ •Vila

Battle of Savo Island
Aug. 8-9, 1942

L

Japanese destroyer Amagiri collides with PT 109
Aug. 2, 1943

•*Munda*

New Georgia

New Georgia Sound (The Slot)

Tulagi (Tulaghi)
Nggela Sule
Florida Is.
Nggela Pile

Malaita

A

Battle of the Eastern Solomons
Aug. 23-25, 1942 🌺

□ *Rendova Harbour (U.S. Naval Base)*

Vangunu

Battle of Kula Gulf
July 5-6, 1943

Rendova
Tetepare

Nggatokae

O

Russell Is.

Savo 🌺

Iron Bottom Sound

N

Battle of Cape Esperance
Nov. 11-12, 1942

□ Henderson Field (U.S. Army Airbase)

Maramasike

Battle of Guadalcanal (Naval)
Nov. 30-Dec. 1, 1942

+ *Mt. Makarakomburu* 8,028 ft (2,447 m)

D

Ulawa

10°S

Battle of Guadalcanal
Aug. 7, 1942-Feb. 9, 1943

Guadalcanal

S

S

Pocklington Reef

E

A

San Cristobal

HIPELAGO

Rossel Island

ula

🌺 **Battle of the Coral Sea**
May 4-8, 1942

Rennell

54°E 156°E 158°E 160°E 162°E

12°S

INTRODUCTION

By Senator Edward M. Kennedy

"I'VE GOT SOMETHING SPECIAL FOR YOU," Jack said with a laugh, "and you get first choice!" There, laid out before my widening eyes, was an extraordinary collection of colorful Solomon Islander war clubs. They were at least four feet long, and as any 11-year-old would be, I was fascinated. An even bigger thrill was that my brother was letting me be the first to pick one out for myself. He had brought the exotic clubs back from the South Pacific, where he had been serving with a PT boat squadron during World War II as the captain of *PT 109*.

Jack had just arrived at our home in Palm Beach, Florida, after the long trip back from the Solomon Islands. On the way, he made a short stop in California to see friends and rest. While there, he met with Mrs. Patrick McMahon, the wife of one of the injured members of his crew. He told her of the night *PT 109* had been rammed by a Japanese destroyer. Two members of his crew were killed, and he and the rest of the crew swam for hours to get to the nearest island. Jack explained how they had survived on coconuts. He said he swam out into the channel at night to try to hail any passing U.S. ship, but none had come. At last, two Solomon Island natives spotted them and that led to their rescue. He was able to reassure Mrs. McMahon that her husband, Patrick, would recover from his injuries and be home soon.

Now, at long last, Jack was home from the war front. We were so glad to see him and so proud of all he had done. He seemed older to me and all grown up, especially in his uniform. I was in awe of his medals, but he valued the friends he'd made in the Navy more. He was my hero, as well as my godfather, and I treasured the special attention he gave to me. It led to one of my most memorable times with him before he left the Navy. In bits and pieces he told us about what he had been through, but I noticed he didn't want to dwell on it. He talked to my father about helping the families of the crew members who had been lost, and made sure assistance was provided over the years.

Because of his experience, Jack was assigned to be an instructor in the PT boat training program in Miami. One day, he asked if I'd like to go for a ride, and the next

thing I knew he had smuggled me aboard one of the PT boats at the base. With the engine roaring, we raced across the waves. It was exhilarating and I loved it! He introduced me to the crew and showed me what it was like to stand at the helm. I also learned you have to watch where you walk when you are around Navy seamen spitting chewing tobacco. I'd never seen anyone do that before. Suddenly I had a shirtfront full of it, which everyone else thought was very funny, but I smelled like tobacco for the rest of the day.

Jack often invited a Navy friend from the base to visit for the weekend. As they were getting ready to go out for the evening, he would ask me to go to the guestroom and deliver a message to his friend. It would all be in Navy slang. Then I would bring a response back to him. It greatly amused him that I carried messages back and forth all weekend and never understood a thing they were saying.

Jack's combat injuries had aggravated his bad back and he was honorably discharged from the Navy. In the spring of 1944, he had to have an operation on his back. He was enormously cheered during his recuperation by visits from his former shipmates. Later, in August, he invited George "Barney" Ross, Leonard Thom, Jim Reed, and Paul "Red" Fay to visit Hyannis Port. It was there that I first met Barney and Red, who became lifelong friends of our family. It was a happy weekend reunion full of spirited football games, laugh-filled outings on our sailboat, and boisterous storytelling. In the evening we all sang Irish songs around the piano with my grandfather, "Honey Fitz."

When Jack decided to enter public service and run for Congress in 1946, it was his Navy friends who were the first to rally to his side, and they helped in his 1960 presidential campaign as well. During the Inaugural Parade, his happiest surprise was seeing his loyal crew waving to him from the deck of a replica of *PT 109*.

Like all the young men who had fought in World War II, my brother's war service was a defining experience in his life. He had coped with fear and danger and death. He had been tested, and he had found within himself the courage to prevail. He had become a leader and had grown and matured. He had been under enemy fire, so he took the hardships and combat of an election campaign in stride. His war record gave him credibility and the respect of his fellow veterans, and the experience of leadership and loss gave him an instant connection to the many families that had suffered too. The Gold Star mothers understood how deeply Jack empathized with them, for they knew our family had lost my brother Joe—and our mother was a Gold Star mother too. They supported Jack as if he were one of their own.

Jack represented a new generation of leadership. His war experience established him as a patriot and prominent figure in his generation. It had taught him the terrible price of war. He believed that serving in Congress, where war is declared, and later in the White House, where it is commanded, he would be in the best possible position to help prevent it.

Not a day goes by that I don't miss him. And I think of *PT 109* each time I sit in my den and look at the family pictures and mementos surrounding me. On the wall is the greatly treasured war club Jack gave me so many years ago. I have always kept it. My mother marked the one I chose with a little white tag. She wrote on it, "To Teddy from Jack." The tag is still there.

August 2002

Previous page: In November 1944 Navy veterans of the South Pacific campaign gathered at the Kennedy home in Hyannis Port, Massachusetts. Linked arm in arm stand Paul "Red" Fay, John F. Kennedy, Leonard Thom, James Reed, George "Barney" Ross, and Bernie Lyons. The boy in front grew up to become Senator Edward M. Kennedy. To his left with arms folded is his cousin Joseph Gargan.

THE
LOST BOAT

CHAPTER ONE

In dress whites, Ensign Kennedy visits home in May or June
of 1942. He had attempted to enlist in the Army but was rejected
for lower back problems. He passed the Navy's physical exam
after remedial exercises and reported for active duty in September
1941. Bored with early paperwork assignments, he angled
for duty in PT boats, famous for their rescue of General Douglas
MacArthur from the Philippines and featured in the 1942 bestseller *They Were Expendable*.

AUGUST 1943-PRESENT

AT THE FAR ENDS OF THE OCEAN, WORLDS ARE COLLIDING.

It's about 2:20 a.m., August 2, 1943, in the western Solomon Islands, 400 miles east of Papua New Guinea and 1,000 miles due north of Brisbane, Australia. Lieutenant (junior grade) John F. Kennedy, all of 26 years old, squints into the blackness while at the wheel of *PT 109*. His mission is to harass an incoming Japanese resupply convoy, and if possible, to sink Japanese barges.

The night is hot and steamy, the sky cloudy and moonless. The surface of the Solomon Sea is a black void, giving no answers or comfort. Dark shapes loom on the impenetrable horizon, wild volcanic islands that are home to peoples only just emerging from headhunting and cannibalism. It is one of the most primitive and isolated places on the planet, a place itself born of tectonic collision as two great plates of the Earth crash up against one another, their impact, except for this island chain, buried under miles of ocean.

Flashes and searchlights flare to the west, perhaps indicating battle. A radio crackles and fades, giving only confusing information. A 37-mm artillery gun is lashed to the bow of *PT 109* with rope because the crew didn't have time to properly bolt it down in advance of the coming engagement, which they had learned about only that day. The crew contains some relative long-timers, who've been in these parts for several months, but mostly these men are green, only recently arrived at the front lines. There is even one man who has asked to come along for the ride but isn't on the crew roster.

They've been doing this night duty for some months, and have adjusted in their own ways to working until dawn and then by day, trying to catch some sleep and do necessary maintenance until they ride out again the next evening. But they haven't been so close to the enemy, and in a position to hit back, until this hot night in August.

They also haven't been in as much danger.

The United States and Japan are locked in a struggle to the death over who will control Asia and the Pacific. After the American debacle at Pearl Harbor and a lightning Japanese advance into Indochina, the Dutch East Indies, British Malaya, and the Philippines, Japan sits astride an empire, euphemistically called the Greater East Asia Co-Prosperity Sphere, that is the largest of all time. The Americans, with British, French, Australian, and other Allied help, have to roll back the Japanese in the Pacific—but without jeopardizing the Allied campaign against Hitler in Europe.

By this dark night the tide is already turning against the Japanese. Swift victories at Pearl Harbor and in China and Southeast Asia have yielded to a stalemate in the Battle of the Coral Sea and utter defeat at Midway and Guadalcanal. Instead of winning and advancing without opposition, Japan is now often losing, and sometimes retreating under cover of darkness.

Thousands of Americans and Japanese have died, in battle upon battle, culminating in the terrible bloodletting at Guadalcanal. First seized by Japan in early 1942, Guadalcanal was invaded by Marines in August to take the new Japanese airfield. That airfield, plus the island's size and position as the southern anchor of the Solomon chain, make it so important for both sides. By February 1943, after months of struggle, Guadalcanal was American. By this night in August 1943, the action has shifted several hundred miles northwest, back up the Solomon chain toward New Guinea. The campaign has focused on islands like Rendova and Kolombangara, only a few miles from where Kennedy waits, as Japan doggedly refuses to concede its holdings.

The Solomon Islands are riven by a water passage known to Americans as the "Slot." Long after sundown, up and down this slot, the Japanese run their desperate supply missions from their huge base at Rabaul, 500 miles northwest, supply runs called the "Tokyo Express." The Express, first created in early 1942 to resupply the huge Japanese presence across northern New Guinea and the Bismarck Archipelago and down the Solomon chain, has tried and failed to save the doomed Japanese on Guadalcanal Island and Tulaghi. By late summer 1943, Japan is using the Express to try to save its

Blasted by 2 torpedoes and 24 shells, the Australian heavy cruiser *Canberra* was mortally wounded in the Battle of Savo Island, Solomon Islands, on the night of August 9-10, 1942. The engagement ended in disaster. Four Allied ships were sunk by Japanese fire; 1,023 men died, and 709 were wounded. *Canberra* was located and photographed by Robert D. Ballard in 1992.

endangered anchorages and troop placements farther southeast in the embattled island chain on New Georgia and Kolombangara, and its airfield at Munda and seaplane base at Gizo. Even though Americans are slowly pushing them back, the Japanese rule the night, by virtue of superior night-vision equipment and night fighting tactics—and the American's fear of sending capital ships forward for lack of air cover. And so the Tokyo Express rolls ahead, despite American opposition.

Lieutenant John Kennedy's orders, like those of the other commanders of PTs and larger ships also deployed from the PT base at Lumbari Island off Rendova, are to attack and harass the Tokyo Express resupplying Japanese installations, tonight the destination being Vila Plantation on Kolombangara Island. He and his young crew of 12 think they are ready for their direct engagement with the Japanese enemy, even if they are handicapped by poor communications, sometimes unreliable equipment and weapons, inexperience, darkness, and confusing information on the enemy's whereabouts.

The PTs are deployed in groups of four. Only one boat in each group has radar equipment, and it is designated the lead boat. When radio silence is imposed, as is the case until the enemy is engaged, American orders are for the backup boats to just follow their leaders—even when the followers don't know why they are moving.

A new black shape looms. Could it be another American PT boat? The other craft can't be identified. Observers on other American PT boats can also see it and the phosphorescent wake thrown up at its bow and stern, indicating high speed. Some of these observers will later tell analysts and historians that they recognized it as a destroyer, and that they saw it headed straight toward *PT 109*. They were certain it was Japanese. Although they claim they radioed repeated warnings to Kennedy's group leader but never got a response, *PT 105* skipper Dick Keresey responds that no such warnings were ever sent. "I heard everything else that was transmitted that night," he says, "and I never heard any such transmissions."

The huge destroyer traveling between 30 and 40 knots plows into the 80-foot wooden-hulled PT boat and tears it into two pieces. Crew members are knocked down or thrown into the water; one is badly burned, and two are killed. Several men who can't swim are faced with the prospect of swimming for their lives. Although they all are wearing life jackets, the life raft had been removed that afternoon to make room for the hastily added gun, which was supposed to supply added firepower against Japanese barges. The volatile PT fuel explodes on the surface of the sea and burns for a short time, illuminating the scene of the wreck. The Japanese destroyer races on toward Rabaul, hardly scratched. Mercifully the burning fuel is pulled away by the suction of the destroyer, so the survivors aren't further tested by fire. The PT's stern has sunk beneath the waves, but the bow remains afloat for a time.

Still no radio communications go out from *PT 109*. From miles away, the other boats watch the burning fuel for a few minutes, but then can see nothing. The sea is black and huge, the night enveloping. The long night slides toward daybreak. The surviving PT boats head back to their tiny base at Lumbari Island. Everyone knows that *PT 109* is gone, and they assume the worst. Maybe everyone on board is dead, lost to fire, to sharks, to drowning. It is not clear if a search is attempted that night. Although Philip Potter, commander of *PT 169*, years later claimed to have looked for survivors, Keresey says that no report at the time of the events mentioned any search.

Search or not, only the crew of *PT 109* knows they are still alive.

The crew is a cross section of American society. Aside from Ivy Leaguer Kennedy, the other officers now adrift are executive officer Ens. Lennie Thom, a blond, Viking-bearded, Ohio-born former college football player and Kennedy's right-hand man, and last-minute ride-along Barney Ross, a beefy, boisterous prankster.

The enlisted men include machinist's mate Patrick McMahon of Los Angeles, at age 37 the oldest crew member and affectionately known as "Pappy"; gunner's mate Charles Harris of Boston, only 20 and most recently a truckdriver back home; machinist's mate Gerard E. Zinser of Belleville, Illinois, a young six-year veteran of the Navy who had trained at the Packard engine plant in Detroit and so has more knowledge of the PT engines than anyone else on board; and radioman John Maguire of Hastings-on-Hudson, New York, 26, who had been enticed into volunteering for PT service by his brother William.

Joining them are machinist's mate William Johnston of Dorchester, Massachusetts, born in Scotland and by the time he is assigned to *PT 109* already weary of the Pacific War; ordnanceman Edgar Mauer of St. Louis, young survivor of one ship sinking and not impressed by his first meeting with Kennedy, reportedly calling him "another 90-day wonder"; torpedoman Ray L. Starkey of Garden Grove, California, going gray at age 29 from a hard life as a former oil field worker and commercial fisherman; motor machinist's mate Harold Marney of Springfield, Massachusetts, only 19 and in the Navy since the tender age of 17; torpedoman Andrew Jackson Kirksey of Macon, Georgia, a quiet Southern man who in recent weeks has been having premonitions of his own death in combat; and Seaman 1st Class Raymond Albert of Cleveland, only 20 and considered cocky by those who know him.

As their shattered boat gradually sinks lower, John F. Kennedy knows his adventure is just beginning, and his life is being rewritten. Any thoughts of doing damage to the Japanese have been replaced by ideas about how best to save his own men. Two are dead. One is badly burned. At least three can't swim. And yet swimming is the only way any of them will make it through the coming days.

Tuesday, May 14, 2002 (en route Hawaii-Australia)

It has always been a dream of mine to search for this important boat. Important not so much in a military sense as in a human and historic sense. A style of leadership was spawned that night, since Kennedy would one day rise to assume the same position that Franklin Roosevelt held in August 1943, and would himself have to deal with issues of war and peace.

John Kennedy defined the ideals of leadership for my generation and beyond—a charismatic athlete-intellectual, a Navy hero, overcoming poor health to achieve great things. His experiences with *PT 109* and World War II were critical in shaping how he would approach future political and military challenges, such as Vietnam, Cuba, and the Soviet Union. Kennedy described his wartime service, including his sinking and rescue, as the single most important experience of his life, one that defined him and his generation. "That war made us," he once wrote to Lady Nancy Astor. "It was and is our single greatest moment. The memory of the war is a key to our characters."

Although I didn't serve in the Solomons during my own Navy career, I did come here in 1991 and 1992 to search for the many wrecks—American, Japanese, and Australian—that still litter Ironbottom Sound, the spot off Guadalcanal where more than 50 ships rest in peace. We found 12 of them, all bigger than *PT 109*, and none of them with wooden hulls. Those wrecks on the sea bottom are timeless memorials to the war that was waged in Guadalcanal and the Solomons.

Those earlier visits gave me a taste and a liking for this remote part of the world. It was the curious destiny of the Solomon Islanders not only to jump instantly from the Stone Age into the Industrial Age, but also to participate in the bloodiest, most devastating chapter of modern warfare. These simple village people who for thousands of years had eked a living from the jungle or from the ocean faced the dangers of modern combat, forced to decide which side in this worldwide struggle they would support.

To show how the smallest event can ripple across history, imagine what would have happened had the islanders who met Kennedy and his crew decided to turn him in to

Bob Ballard checks the remotely operated vehicle (ROV) named *Little Hercules.*
Its mission is to capture high-definition television (HDTV) images of any remains
of *PT 109* that might be visible on the sea bottom. Two other underwater sleds
complete the team. *Echo* uses side-scan sonar to locate potential targets. *Argus*
travels with *Little Herc,* to provide lighting for the HDTV camera.

Pages 16-17: A PT boat's (here depicted as *109*) nightmare was to be caught in close
formation in daylight by attacking Japanese planes. Antiaircraft firepower included
four .50-caliber machine guns and two 20-mm cannons, but the best defense was the
boats' evasive maneuvers; they could travel at speeds of up to 45 knots (52 mph).

the Japanese? What if they had killed the Americans on the spot? A small event in the context of the war, but a crucial one in world history. Without John F. Kennedy, there would have been no Peace Corps, no New Frontier, and maybe no NASA program to put men on the moon.

I was drawn to the search for other reasons as well. Although I'm not a native-born New Englander as Kennedy was, I now make my home in Lyme, Connecticut, about an hour's drive from Melville, Rhode Island, where Kennedy learned to pilot a PT boat, and another few hours from Hyannis Port out on Cape Cod, where he and his siblings learned to sail and where I lived for 30 years. I've been up and down that coast many times, by land and by sea, whether on business or vacation, or when I was once on the faculty at the Woods Hole Oceanographic Institution in Woods Hole, Massachusetts. The Kennedys and their legacy are part of that seascape.

And I've always been interested in the controversial legacy of the PT boats themselves. During World War II, the legendary exploits of the small, fast, unarmored, and thus vulnerable boats kept American spirits up, even though the story behind the scenes was not so glamorous. As I would learn, in the grim months right after Pearl Harbor, the dashing boats and their daredevil crews did more to boost American spirits than to stop the Japanese. What was the real story behind the PT saga, especially in those days of 1943?

The search for *PT 109* was also a technical challenge. Sixty years after Kennedy's collision, finding one or two sections of wooden hull at 1,300 feet in a tricky tropical sea was no mean feat. Add to that the turbulent volcanic terrain and seafloor in the area, including active seamounts that occasionally rise steaming above the waves, only to vanish again. And the big volcanoes like Kolombangara and Savo that are not extinct, only quiescent.

In May 2002, I returned to where it all began—to the place where one young American came face to face with death. He wasn't the greatest PT strategist; he never had the time to become one. But he risked his life to save his men.

OUR SEARCH VESSEL WILL BE *GRAYSCOUT*, up from Gladstone in Australia. She's normally a charter, used for deep-sea fishing and wreck diving on Australia's Great Barrier Reef. This time out, she's looking for a World War II heirloom. *Grayscout* is 112 feet long, with a 28-foot beam. She has two GM diesel engines and 8 twin berth cabins that are alleged to be air-conditioned.

The co-stars of the expedition are sonar sled *Echo*, imaging sled *Argus*, and ROV (remotely operated vehicle) *Little Herc*. *Echo*'s role is to do side-scan sonar of our search area—in other words, to provide a rough sonar map of the ocean bottom and of potential targets. Once we analyze the targets, we zoom in more closely with our underwater video system, made up of *Argus* and *Little Herc*. All three vehicles are based at the Institute for Exploration in Mystic, Connecticut, and were designed and built by my team headed by Jim Newman.

Argus is a towed imaging vehicle—an underwater "mother sub" with lights and video camera connected by cable to the surface ship. She's been to places like the Black Sea and my last World War II search, at Pearl Harbor in 2000, when I went looking for one of the five Japanese mini-subs that tried to sneak into Pearl Harbor.

Little Herc is in turn connected to *Argus,* and she is equipped with one of the finest underwater high-definition video cameras in the world. The picture she delivers is the best you can get, fine-tuned by our contractor, Jay Minkin. While *Argus* provides the underwater lighting, *Little Herc* does the serious underwater videotaping of potential targets.

My crew of nine consists mostly of familiar faces. Aside from the additional Aussie crew of seven assigned to *Grayscout,* my chief operating officer is the ever capable and unflappable Cathy Offinger, back fresh from a scouting expedition in South America. Cathy makes it all happen, from finding the search vessel and getting submersibles from Connecticut or wherever to the Solomons, on time and intact, to finding hotels and flights and grub. That was no small task in the Solomons, which though they are not as remote as in 1943, still take about 34 hours of air travel to reach from the U.S. Co-watch leader Dwight Coleman will be with us, together with engineers Jim Newman and Dave Wright, mechanical engineer Todd Gregory, deck engineer Mark DeRoche, Dave Lovalvo, and Jay Minkin, master of images.

We're using only one search vessel, *Grayscout,* and several small launches to move us back and forth from the ship to our shore base on the island of Gizo.

Those are the players. But how to find this long-lost wreck? Although my earlier searches had often been aided by precise coordinates calculated by sophisticated instruments on large vessels, the PT boats didn't have the instrumentation, and their captains were not as diligent in recording times and places at which events occurred. As I was to learn, what reports they did prepare were usually done on shore, after an action, based on recollections. PT logs were usually pretty sketchy affairs. In the case of *PT 109,* the logbook had gone down with the ship.

On the other hand, we are favored in that we'll be working in a fairly limited area, only about 35 square miles, in the Blackett Strait a few miles off Gizo. We have witnesses from other boats, such as *PT 162*'s John Lowery and *PT 169*'s Philip Potter. Although the sinking hull of *PT 109* had drifted some after the collision, we are surveying a search area of some square miles, not hundreds of square miles in mid-ocean.

Making things tougher is our brief expedition time—a little over a week, necessitated by *Grayscout*'s tight charter schedule. This means we don't have the luxury of sifting through lots of extraneous data. We will need to home in fast. And then there are the quirks of sea changes and bottom currents and ocean chemistry to factor in. Just what do we hope to find? Will it be intact enough to be recognized for what it is? Or have war and the passage of time turned *PT 109* into flotsam and jetsam?

THE YOUNG

CHAPTER TWO

SKIPPER

Hyannisport 1934

No war clouds yet dim a gathering of Kennedy children in 1934. Eunice, at left, stands next to big brother, Joseph Jr., and little brother, Robert. John holds Jean aloft next to Rosemary, and Patricia sits in front. John, called Jack, has another year to go in prep school, where he makes only fair grades, gets into school-boy scrapes, and is voted "Most Likely to Succeed."

1917–1943

CERTAIN QUESTIONS MUST BE ANSWERED to give a fuller accounting of this seemingly small historical event long ago. The most important one is: Just who was this young man who captained a PT boat in a terrible war at the far ends of the ocean?

John Kennedy was both blessed and burdened. Blessed with intelligence, looks, friends, wealth, and power. And burdened by bad health, a family legacy to uphold, and the need to prove wrong the prejudice against his faith.

Left now are television documentaries and interviews, photos of another time. Looking there, trying to find the nature of the man in those imperfect records from his youth before the war, you see his good looks, his smile. You see the picture-perfect settings of privilege, the parties at Hyannis Port and Palm Beach, the family gatherings. He projects health and athleticism, even if those qualities don't come naturally to him. Reading the historians, you get a sense of a charming young man most interested in having fun.

A handful of surviving veterans would go on to become President of the United States—Dwight D. Eisenhower, John F. Kennedy, Lyndon B. Johnson, Richard M. Nixon, George H. W. Bush. Each went through the common crucible of World War II. But among this small fraternity, Kennedy stands out. In an electronic age that was able to magnify its leaders a thousand times and into every corner of every home, Kennedy excited the American people as no other.

Why was that? Although some of his charisma was unique, another part of it derived from his family. While John Kennedy was an individual with his own destiny, personality, and ambitions, he was also a member of a family that had, and continues to have, exceptional impact on national affairs. It also had powerful impact on those born into it, and gave them certain advantages.

The origins of the Kennedys are in Ireland, which their ancestor, Patrick Kennedy, was driven by poverty to leave in 1848 to resettle in Boston. Patrick died during a Boston cholera epidemic in 1858, but by then he had managed to move his heirs—five children and a wife—out of poverty, leaving them a small store.

The Kennedys were Catholic, and the American elite of the 19th century had inherited the British ambivalence about Catholicism that goes all the way back to King Henry VIII. The Kennedys' Catholicism made them outsiders.

The Kennedys became associated with politics. Although the Mellons, Carnegies, and Vanderbilts never engaged directly in government affairs, the Kennedys, like the patrician Rockefellers, Roosevelts, and Bushes, gravitated naturally to government. By 1887, immigrant patriarch Patrick Kennedy's youngest son, P. J., was a prosperous merchant and a Democratic member of the Massachusetts House of Representatives, and was married to the sister of Boston's mayor. Even with those improving fortunes, however, the Kennedys found their social rise hampered by Protestant Brahmins who dominated society. The Brahmins took their unofficial name from the hereditary ruling caste in India because their own social position seemed as elevated and impenetrable. They barely tolerated newcomers like the Kennedys, and certainly never invited them into their inner circles.

P. J.'s firstborn was Joseph P. Kennedy, who early on acquired his parents' resentment of exclusion and who also absorbed the family drive to succeed through hard work. Although the Kennedy family was now financially comfortable, Joe was still required to make his own way with assorted neighborhood business ventures. Joseph's mother broke tradition with Irish Catholic schooling and sent him to Boston Latin School, where he made lifelong friends outside the Irish community and emerged an outstanding baseball player. He then attended Harvard, the seat of Protestant elitism, and achieved an average academic record coupled with athletic and social achievement. He went into banking and became president of his father's Columbia Trust Company while still in his 20s. Joe and Rose Fitzgerald were married by a cardinal in 1914, and two great Boston families were united.

Rose's father, John Fitzgerald, better known as "Honey Fitz," was an enthusiastic Democratic politician who served several terms as Mayor of Boston. He'd also served

The Kennedy children visit Osterville Beach, from left: Rosemary, Jack, Eunice, Joe Jr., and Kathleen (above). Though sickly as a child, Jack was swimming at age eight. As a newly minted Harvard graduate (below), Kennedy had already won two significant sailing regattas, and small-boat-handling experience was just what the Navy was to look for in PT boat skippers.

in the Massachusetts House when P. J. Kennedy was there. Fitzgerald astonished Boston with his grueling campaign and governing style, attending as many as six events in one evening and delivering up to ten speeches a night. As with many big-city politicians of the time, although he sometimes ran as an enemy of the "big bosses" and backroom boys, periodic accusations of scandal and corruption tainted his record.

From his youthful post as president of Columbia Trust, Joseph P. Kennedy went on to run a huge shipyard in Quincy, then later took a pay cut to become manager of the stock department of a major brokerage, Hayden, Stone and Company, where the company's establishment prestige and connections would serve him well. His growing responsibility and visibility put him in touch with national figures like Franklin D. Roosevelt, then Assistant Secretary of the Navy. Skillful Wall Street trading made Kennedy a multimillionaire by the mid-1920s. He also entered the film business in that decade, helping create the emerging entertainment conglomerate RKO and growing even wealthier in the process. On the eve of the 1929 stock market crash, he sold short on his investments and made millions more—so much that it's estimated that by 1930 Joe Kennedy was worth more than $100 million.

Family members, friends, and even Rose herself say that it is the figure of Joseph P. Kennedy that defined the entire family. He more than anyone created the Kennedy ethos of achievement, glamour, power, and charisma; he designed, nurtured, and guided this ethos until his death. A major investor in Hollywood, he learned how to put the machinery of image making at the service of political power.

Second son John Fitzgerald Kennedy was born on May 29, 1917, in Brookline, Massachusetts. He was preceded by sister Rosemary and brother Joseph Jr. According to all accounts, it was understood from the earliest years that Joe Jr. was the anointed heir, though the father took an intense interest in the progress and affairs of all his children. Joseph Jr. favored his father in personality, looks, and intelligence, and as the patriarchs's career took him away for long stretches in Hollywood and elsewhere, Joe Jr. often served as surrogate father for the younger children.

John was different. From birth into adulthood he was dogged by poor health and a vulnerable constitution. He suffered bouts of diphtheria, scarlet fever, jaundice, and possibly hepatitis. He often succumbed to bronchitis and pneumonia, sometimes landing in a hospital or infirmary. His vision was poor, and he seems to have had a congenital back problem that was later blamed on football or war injury. To those who knew him, the aching back was a fixture in his life. Later accounts of his health allege Addison's disease. Whatever the different afflictions, John F. Kennedy spent many days in pain.

Though John was not a natural athlete or physically strong, he was still a fierce competitor. The family football tradition at Hyannis was serious business; winning was everything, and John participated vigorously. During summers, father Joseph brought in trainers and sports coaches, and the boys were encouraged to sail. Joe Jr. apparently taught John how to sail, and so was begun JFK's lifelong association with the sea.

But Kennedy's physical type naturally inclined him more to intellectual than physical pursuits in his early years. He was an avid reader, which opened him up to a broad variety of subjects. But the reading didn't translate into stellar academics. Throughout boarding school at Dexter and prep school at Choate his grades were middling, peppered with a D or two, and he was often chided for tardiness and sloppiness. At Choate he had to follow in the footsteps of elder brother Joe, who was finishing out his scholastic career as the campus star. Joe excelled in his studies, campus politics, and social life. In contrast, John did well only in the social arena, where he was more known for his sense of humor than his leadership. He initially chose Princeton for college rather than Harvard, as his father wanted, but poor health undermined his studies and he left after the first semester, suffering from apparent hepatitis. After a stint in Arizona to get his health back, he yielded to his father's wishes and enrolled at Harvard the next year.

At Harvard he tried out for freshman football and swimming, and when his health allowed, he competed at the margins. His friends tended to be athletes, and although he couldn't beat them, they liked him because of his sense of humor and charm. For his part, he loved to be around them. Some have speculated that this was his attempt to keep up with athlete brother Joe; if he couldn't beat them, he would certainly join them.

As John was entering Harvard, his father was reaching the peak of his political career, although he hoped it was the beginning of even bigger things. Although the patriarch hadn't chosen the route of electoral competition like his own father, P. J., he managed to advance through appointment. First he ran Franklin Roosevelt's Securities and Exchange Commission, where his Wall Street trading experience served him well. By late 1937, Kennedy was named ambassador to the Court of St. James's in London, not Cabinet secretary level as he might have preferred, but the most coveted diplomatic assignment.

The elder Kennedy arrived as Europe was sliding toward war, giving his son John a unique entrée into a society looking at the abyss. John traveled often to London, Europe, and elsewhere while his father was ambassador. He even undertook a semester away from Harvard in spring 1939 and carried out some personal "research missions"

In the United Kingdom, U.S. Ambassador to the Court of St. James's, Joseph P. Kennedy and family turn out on the Fourth of July, 1938, at the embassy residence (from left: Eunice, Jack, Rosemary, Jean, Joseph P. Kennedy, Edward, Rose, Joe Jr., Patricia, Robert, and Kathleen). By seeing the growing turmoil in Europe firsthand, Jack Kennedy brought a deeper understanding and interest to his studies at Harvard.

in places like Prague, Berlin, Moscow, and Palestine that became the basis for letters to his father about the gathering storm and world affairs. JFK was in Berlin just before Hitler invaded Poland.

The elder Kennedy is said to have been considering a future run for President himself. But that ambition was thwarted. Ambassador Kennedy was a devout isolationist, and though the United States remained officially neutral until the shock of Pearl Harbor, President Roosevelt—and much of the country—was moving in another direction, preparing for what seemed like an inevitable war with fascism. The enormously popular Roosevelt reportedly had concluded by 1940 that he and Joseph Kennedy just didn't share the same worldview, and so his once rising star was dimmed. Whether Kennedy's isolationism was deeply held or the result of fears that his precious sons would be sent into war is not known. Whatever the roots of his dissent, by late 1940 Ambassador Kennedy had decided to resign.

The elder Kennedy worked to avoid war, and his sons were publicly loyal to his views. Joe led an isolationist student group at Harvard Law School, while John published his Harvard thesis on England's reluctance to oppose Hitler as *Why England Slept,* which was well reviewed and attracted some attention. The book ran counter to the prevailing view that Neville Chamberlain's appeasement of Hitler at Munich diverged from British national wishes. Kennedy argued that Britain was deeply wary of war, and so appeasement had deeper roots than one prime minister.

As Joe excelled at Harvard Law School, John relaxed after graduation from college. More than anything else in that period, he liked to enjoy life. Many biographers say girls were a major pursuit, and while he chased them as sport, he was good-natured enough to leave most of them with fond memories. He was interested in dressing well and partying. He was constantly invited to visit family friends or to attend house parties in Florida or North Carolina or California.

But he knew he had to make a stab at building a career. He audited a semester at Stanford Business School, then dropped out when plagued by ulcers. The first half of 1941 was spent traveling in the U.S. and South America. By that summer, the rising drumbeat of war had seriously undercut the isolationist viewpoint, and the Kennedy boys began to think about serving their country. For John, it would be the war, not Harvard or a traditional career, that would propel him in a more serious direction. Joe Jr. enrolled as a naval aviator, while JFK became an ensign in the Naval Reserve and briefly outranked his older, more accomplished brother, who had not yet finished his officer training.

Returning from Europe in September 1938 aboard Germany's SS *Bremen,* which had taken the blue ribbon for fastest transatlantic crossing from Britain, a smartly turned out Jack Kennedy heads for college. A version of his senior thesis, published commercially as *Why England Slept,* brought celebrity.

At Harvard, backstroker Kennedy posted no major records, but
the training helped strengthen his chronically weak back. It
also helped prepare him for the swim for his life and the lives
of his crew in the far more dangerous waters of Blackett Strait.

John was lucky to pass his physical, as his bad back was a liability. His back was so bad by this point that surgery was proposed, but he sidestepped it by wearing a corset and sleeping on a board. His father's political influence helped him to get an assignment at the Office of Naval Intelligence (ONI) in Washington, where he helped prepare the daily intelligence briefings for the Navy brass based on cables and clips from around the world. He saw the first reports about the surprise Japanese attack on Pearl Harbor and the terrible destruction it had wrought during his time at ONI.

In early 1942, Kennedy was transferred to an ONI office in Charleston, South Carolina, a move that disappointed him. To go from the center of action in Washington to a minor outpost was galling, even as the real action was thousands of miles away in Europe and the Pacific. His complaining resulted in reassignment to Northwestern University and Midshipmen's School in Chicago in preparation for active duty. While there, he volunteered for PT boat duty.

A PT boat command looked like a natural for him, if he could stand the rigorous conditions and the constant pounding at high speed across ocean waves that would take a toll on any back. After all, he had grown up on the water. PT boats were exciting and dashing, giving their crews much more autonomy, adventure, and responsibility than service on a destroyer or aircraft carrier. Such big ships were crewed by hundreds or thousands of men, turning them as much as anything into giant floating bureaucracies. Dress codes and saluting officers, for example, weren't required on PT boats. This sat well with Kennedy, who had had brushes with authority, whether with his father or Joe Jr., as well as in boarding school and college and with his early Navy superiors; although he wasn't a dogged rebel, he certainly wasn't spit and polish either.

The PTs tended to attract his type—athletes, Ivy Leaguers, college men, and others who wanted a challenge and didn't relish trying to fit themselves into the more controlled society in a bigger organization. Rather than wait decades to command a large craft, PT men could earn a command, as did Jack, when they were still green.

Of the 720 PT boats built in World War II, more than 500 of these boats were put into service and 70 of them were lost. They are best remembered for their fight against Japan in the Solomons, New Guinea, and the Philippines, but they also took on the Germans in the North Atlantic and the Mediterranean. Combat eliminated up to 30 of them, while the others were victims of friendly fire, accidents, or the inevitable deterioration caused by salt water, dry rot, wood borers and other nautical hazards.

The dark side of PTs was that they were dangerous, lacking enough armor to shield the thousands of gallons of aviation fuel from explosion when hit by a Japanese bullet or

Promoted to Lieutenant, Junior Grade, Kennedy (top row, seventh from right) completed the course at the Motor Torpedo Boat Training Center at Melville, Rhode Island. After a stint as a PT instructor and orders for Panama, he was assigned to combat posting at Tulaghi Island in the Solomons as captain of *PT 109*.

shell. And as later Navy investigations revealed, at the beginning of the war they were embarrassingly ineffective, hardly ever hitting, much less sinking, Japanese ships with their torpedoes. This was due in large part to the fact that outdated WWI-era American Mark VIII torpedoes were too slow to catch up with a Japanese destroyer moving at full speed. To get in close enough to their targets to make up for that deficit, lightly armored PT boats ran the risk of taking a fatal hit from Japanese guns and bombs. Yet in the popular mind and for the desperate American propaganda machine after the debacle of Pearl Harbor, PTs looked like winners. And the Allies badly needed winners to boost morale.

From Northwestern, Kennedy went on to PT training school, in Melville, Rhode Island, in autumn 1942. Although Melville, just above Newport and not far from the family compound in Hyannis Port, was beautiful in summer, in cold weather it became bitterly cold and windy, prone to northeasters, driving rain, and heavy seas.

One of his fellow officers at Melville that winter was his future friend and the future Navy Undersecretary Paul "Red" Fay. Their first meeting was inauspicious. Fay, who like many PT recruits was an athlete and had played football at Stanford, organized a touch football game one afternoon. A thin young man in an inside-out sweater asked to join. Fay, thinking he was a local high school kid, sent him to the other team. But the young man played so hard that Fay took note of him. Imagine his surprise when he reported for duty the next morning to find that the supposed high school kid was Ensign Kennedy, his instructor.

The relationship initially went downhill when Fay, disobeying orders, jumped aboard a new 80-foot PT instead of the older and slower 77-footer he'd been assigned to. When he finished his morning's training he was told to report to Kennedy.

"Kennedy chewed me out," recalls Fay. "He said if everybody in the Navy acted like me, the Japanese would be in Times Square in a month." Threatened with expulsion, Fay begged for another chance.

Appointment as an instructor might have seemed an honor to others, but John wanted to go to war, not teach others bound for the action. He constantly complained about being "shafted" by the Navy, and for a while Kennedy had the dubious nickname of "Shafty."

Shafting didn't sit well with him, even if it was unintended, or intended to save his bad back from the pain of combat duty. Again he communicated his disappointment. And again he was rescued from the back bench, in early 1943 obtaining a transfer to PT combat duty in the Solomon Islands—the very front lines in the American attempt to roll back the Japanese juggernaut.

Pulling friendly rank on brother Joe Jr., of whom great things were
expected—perhaps to serve as the first Roman Catholic President
of the United States—Jack pushes his arm forward to show the two
stripes, which outrank his brother's one. Joe, a Navy pilot, was
killed on a secret mission in August 1944.

Rumpled khaki fatigues rather than dress whites
or navy blues were the uniform of the day in
1943 on Tulaghi Island, the Solomons. Kennedy
picked the crew for his new command and
with them set to scraping, cleaning, and painting
109 and running routine patrols.

With this turn of fortune, Kennedy's Navy career also once again leapfrogged past that of his brother Joe Jr., who was temporarily cooling his heels in Puerto Rico, flying Caribbean antisubmarine duty that didn't turn up many Axis submarines. At the end of March 1943, after a brief stint in Panama, John Kennedy took a converted French steamer turned troopship, the USS *Rochambeau*, to the western Pacific jumping-off point of Espiritu Santo in the New Hebrides, and from there he went on to the American base at Tulaghi in the Solomons.

While en route to his assignment Kennedy tasted his first combat. Japanese aircraft staged a vicious daytime raid on the convoy, and bombs and shrapnel exploded around him for the first time. American troops gave as good as they got, and some Japanese aircraft were shot down. As the ship plowed ahead, it approached a Japanese survivor floating in the water. The ship moved closer to perhaps offer assistance or take him prisoner, but the Japanese began firing madly up at the Americans on the ship's rail. He was answered with a volley of return fire and dropped below the surface of the waves.

After this violent welcome, Kennedy's life settled into a deceptively tranquil routine once at Tulaghi. Although the occasional Japanese air attack was to be expected, the war had entered a temporary lull as both the U.S. and Japan regrouped after the five-month bloodletting of the land battle for Guadalcanal and the half dozen related naval battles in the surrounding seas. Tens of thousands of men had died and dozens of ships had gone to the bottom, and even though the U.S. had turned the tide, the logistics support necessary to capitalize on its gains was not yet in place.

Kennedy wrote his family on May 14, 1943, from Tulaghi:

Dear Dad and Mother:

Received your letter today and was glad to hear everyone was well. Things are still about the same here. We had a raid today but on the whole it's slacked up over the last weeks. I guess it will be more or less routine for another while. Going out every other night for patrol. On good nights it's beautiful—the water is amazingly phosphorescent—flying fish which shine like lights are zooming around and you usually get two or three porpoises who lodge right under the bow and no matter how fast the boat goes, keep just about six inches ahead of the boat. It's been good training. I have an entirely new crew and when the showdown comes I'd like to be confident they know the difference between firing a gun and winding their watch.

Have a lot of natives around and am getting hold of some grass skirts, war clubs, etc. We had one in today who told us about the last man he ate. "Him Jap him are good."

All they seem to want is a pipe and will give you canes, pineapples, anything,
including a wife. They're smartening up lately. When the British were here they had
them working for 17 cents a day but we treat them a heck of a lot better. "English
we no like" is their summating of the British Empire.

I was interested in what you said about MacArthur's popularity. Here he has none—
is, in fact, very, very unpopular. His nick-name is "Dug-out Doug" which seems to
date back to the first invasion of Guadalcanal. The Army was supposed to come in and
relieve the Marines after the beach-head had been established. In ninety-three days
no Army. Rightly or wrongly (probably wrongly) MacArthur is blamed. He is said to
have refused to send the Army in—"He sat down in his dug-out in Australia," (I am
quoting all Navy and Marine personnel) and let the Marines take it.

What actually happened seems to have been that the Navy's hand was forced due
to the speed with which the Japs were building Henderson Field so they just moved
in ready or not. The marines took a terrific beating but gave it back. At the end
the Jap's wouldn't ever surrender til they had found out whether the Americans were
Marines or the Army, if Marines they didn't surrender as the Marines weren't taking
prisoners. In regards to MacArthur, there is no doubt that as men start to come
back that "Dug-Out-Doug" will spread—and I think would probably kill him off. No
one out here has the slightest interest in politics—they just want to get home, morning,
noon and night. They wouldn't give a damn whether they could vote or not and
would probably vote for Roosevelt just because they knew his name.

As far as the length of the war, I don't see how it can stop in less than three years,
but I'm sure we can lick them eventually. Our stuff is better, our pilots and planes
are—everything considered—way ahead of theirs and our resources inexhaustible
though this island to island stuff isn't the answer. If they do that the motto out
here "The Golden Gate by '48" won't even come true. A great hold-up seems to me to
be the lackadaisical way they handle the unloading of ships. They sit in ports out
here weeks at a time while they try to get enough Higgins boats to unload them. They
ought to build their docks the first thing. They're losing ships, in effect, by what seems
from the outside to be just inertia up high. Don't let any one sell the idea that everyone
out here is hustling with the old American energy. They may be ready to give their
blood but not their sweat, if they can help it, and usually they fix it so they can help it.
They have brought back a lot of old Captains and Commanders from retirement and
stuck them in as the heads of these ports and they give the impression of their brains
being in their tails, as Honey Fitz would say. The ship I arrived on—no one in the port

In place on deck, *PT 109* was secured aboard Liberty Ship S.S.
Joseph Stanton for a 1942 passage from Norfolk Navy Yard
to the Pacific. Torpedo tubes are already mounted, as are the six
canisters on the stern, mufflers for routing exhaust underwater
at low speeds for silent running.

Shallow-draft, wooden PTs were speedy platforms on which to mount weapons.
Early versions of such craft served Italian and British forces in World War I,
and one of the early PT prototypes was designed by Englishman Hubert Scott-
Paine, assisted by T. E. Shaw, an alias of T. E. Lawrence, of Arabia.

had the slightest idea it was coming. It had hundreds of men and it sat in the harbor for two weeks while signals were being exchanged. The one man, though, who has everyone's confidence is Halsey, he rates at the very top.

As far as Joe wanting to get out here, I know it is futile to say so, but if I were he I would take as much time about it as I could. He is coming out eventually and will be here for a sufficiency and he will want to be back the day after he arrives, if he runs true to the form of every one else.

As regards Bobby, he ought to do what he wants. You can't estimate risks, some cooks are in more danger out here than a lot of flyers.

Was very interested to hear what your plans were and the situation at home. Let us know the latest dope whenever you can. Whatever happened to Timulty? Jerry O'Leary is out here to the South of where I am, but I hope he will get here some one of these days. He has command of a 150-foot supply boat.

Feeling O.K. The back has really acted amazingly well and gives me scarcely no trouble and in general feels pretty good. Good bunch out here, so all in all it isn't too bad, but when l was speaking about the people who would just as soon be home I didn't mean to use "They"—I meant "WE."

48

I figure should be back within a year though, but brother from then on it's going to take an act of Congress to move me, but I guess that act has already been passed—if it hasn't it will be.

My love to every one.

(signed) Jack

P.S. Mother: Got to church Easter. They had it in a native hut and aside from having a condition read [sic] *"Enemy aircraft in the vicinity" it went on as well as St. Pat's.*

P.P.S. Airmail is better than V-Mail.

Dear Eunice—

Please note that Jack says to send his mail regular airmail instead of V-Mail. In his last letter he also mentioned this and said the service was faster.

Paul Murphy (from Ambassador Kennedy's office)

At Tulaghi, John was quickly given command of *PT 109* because hands were short and he was eager. He found *109* a neglected, decaying craft riddled with cockroaches and peeling paint due to being unmanned for a time in a jungle

Headquarters for the PT squadrons sat in the mud of Florida Island in a compound of huts, a repair shop, and marine railway. Housing was basic. Chow was Spam any way you like it, with occasional scrounged eggs, bread, and powdered ice cream, liberated beer, and 190-proof torpedo fluid (with juice).

With hulls almost flat aft, PT boats planed easily in calm water but pounded hard in heavy seas. Early tactics called for a slow night approach to an enemy vessel and firing torpedoes when inside a 1,000-yard range, then making a fast getaway. Targets were often motorized barges resupplying Japanese troops.

environment, but he and his growing crew worked to make it shipshape. They succeeded well enough for the ship to do training patrols by May, though the real action was still farther west.

Though rudimentary, life at Tulaghi could have been worse. When off duty, John wrote and received letters from home, and he read books under a mosquito net while the other officers played cards. By day everyone worked on the boats and slept. The training runs, like the real thing in the Slot, were done at night, because night was when the Tokyo Express ran. At the end of May, now with a crew of ten, *PT 109* was assigned to the Russell Islands, not yet in the thick of the action but a lot closer.

Other future leaders and future Kennedy associates also moving into the Russells area included intelligence officer Byron "Whizzer" White, a Rhodes Scholar and future Supreme Court justice, future Lieutenant Governor of Virginia and Ambassador to Australia William Battle, and of course Red Fay.

Most important to Kennedy's immediate future was *PT 105* skipper Dick Keresey, Dartmouth graduate and young New York attorney prior to joining the Navy. He had volunteered to avoid the draft. "The main reason I chose the Navy was because I didn't want to sleep on the ground like the Army did," he joked. Besides having an Ivy League background, Keresy was Irish Catholic and a Democrat like Kennedy. The two arrived in Tulaghi at roughly the same time in the spring of 1943.

"If it's possible to be depressed when you're 26, I was depressed the first week I was in Tulaghi," recalls Keresey. "The idea of spending five years or the rest of my life fighting in that place really wore me down. We were part of a group of 20 officers who knew each other. The first time Kennedy made an impression on me was when he hauled my PT boat off a reef. The Solomons are made up of hundreds of islands and reefs, and I managed to run aground on the only reef that was marked. That was the first time I really saw Kennedy in action as a PT skipper. Understand, PTs are very hard to pilot. He towed me alongside after he pulled me off the reef, and he had to pull a very difficult maneuver to get me to drydock; he had to release me so that I had enough forward speed so that I could drift into the dock, because my propellers were shot. I don't know how he did it, but he got me in there.

"After that I'd see him around Tulaghi, along with my friends John Iles and Al Webb. I saw the most of him in the Russell Islands. Kennedy was a real asset. All we had for recreation was conversation, and he was a good talker. He was very amusing, but unlike what some of the others say about him, he didn't strike me as intellectual.

"In the Russell Islands we lived in a plantation house. It looked out over Starlight Channel, and James Michener visited there. The place was so beautiful that I think it inspired him to write *Tales of the South Pacific.*

"I'll be honest, though; I never thought Kennedy would be President. I didn't even know he was an ambassador's son."

In mid-July, Kennedy and crew were assigned to Lumbari Island, more directly adjacent to the Slot. He passed into the command of one Thomas Warfield, an older, very serious career Navy PT squadron commander. He was the type that seemed destined to clash with the more undisciplined Jack Kennedy.

Warfield was dismayed at the odd collection of personalities he had to deal with, because he believed many of them didn't have the mettle and discipline and experience of Navy lifers. Warfield reportedly arrived in the area demanding to have a brig built to detain a Navy man who had riled him on the voyage out, but in the fairly loose atmosphere of Tulaghi his superiors turned him down.

Dick Keresey holds Warfield and others at the PT command center on Rendova responsible for Kennedy's sinking and abandonment. "When Warfield ordered all boats that had fired their torpedoes back to base that night, he also took out the only four boats that had radar. So Kennedy and the rest of us were left behind without radar. We were blind. The base command was filled with people who didn't know their ass from their elbows about what was happening in the field."

It was Jack's good fortune that one of his silliest mistakes occurred in the Russells, just before he joined Warfield's squadron. On the night training runs, Kennedy and the others normally spent 12 to 13 hours out on the water, roughly from dusk to dawn. Toward daybreak, all the boats would race back to be refueled at the Russell docks, the prize being the honor of getting to bed first while the others cooled their heels in the fueling line. One July morning, Kennedy (who reportedly took to the return race with gusto) kept at full throttle well up into the harbor, assuming that he would reverse props at the last minute to cut his speed. But that particular morning, the balky *109* engines stalled when he threw them into reverse, allowing the boat to slam at considerable speed into the dock, and throwing assorted men and tools into the water.

In the rawer and more stressful environment of Lumbari, the gravity of approaching combat must have settled in on Kennedy and his counterparts. Though *PT 109* through good fortune avoided the worst actions in the run-up to August 1, it was repeatedly in danger. The most serious was on the night of July 19-20, when *109* and two other boats patrolled off Gizo in search of Japanese barges.

The night was quiet until a Japanese float plane flew by, an omen of what was to come. Then came radio reports of a full-bore Tokyo Express barreling down from the north, headed through nearby Vella Gulf. The PTs were ordered to pursue, and in doing so passed the forbidding 8° south latitude line, normally their boundary of operations. U.S. ships were banned north of that line so that American aircraft could bomb ships at will on the assumption that they were Japanese.

In fact, misidentification of ships and aircraft was a terrible problem for both sides during World War II. The PTs were repeatedly guilty of it. In June, a squadron commanded by Robert Kelly had attacked Admiral Turner's flagship *McCawley*, a disabled transport that was being towed to repair. Kelly's boat attacked the ship so vigorously by torpedo that it was sunk. A month later, just a few nights before Kennedy's collision, 12 PT boats under Warfield's command mistakenly attacked a convoy directed by Adm. William "Bull" Halsey. On July 20, the responsibility for avoiding mistakes fell on *PT 109*. It was ordered, with two other boats, to proceed into the dangerous northern zone to find and attack the Tokyo Express.

They searched in vain for hours and decided to give up the chase. As they did, a Japanese plane illuminated the Americans' boats with a powerful flare and began to attack. Kennedy pushed the throttle all the way forward as Japanese bombs just missed the boat, though shrapnel took down two of his crew. Some accounts say that Kennedy surrendered the helm to Lennie Thom and attended to the wounded.

The only clue as to what was on Kennedy's mind as his crew made their predawn rush back to Lumbari is a letter he wrote home afterward. Aside from two men wounded, Kennedy wrote his family that he had to contend with the fears and premonitions of another mate, Andrew Kirksey, who was badly shaken by the experience and feared his death. Kennedy took that fear seriously and set about trying to have him replaced.

Kennedy had other engagements on succeeding nights, not so direct but tinged with the potential for disaster. *PT 109* was in or near action at least four or five times. On the night of July 19-20, two PTs fired on an American B-25 and brought it down. It crashed onto one of the boats and destroyed it, killing several men in the process. Some accounts say this screwup so enraged Admiral Halsey that he became determined to get the PTs under his central command.

On July 30-31, *PT 109* itself suffered a setback. On its way out to patrol stations, its rudder stuck and it had to return to base for repairs. Even as Halsey and others were concluding that PTs did not function as well as torpedo boats, the commanders decided

As a consequence of the Allies' strategy of island-hopping, Japanese soldiers were in poor health and near starvation; some POWs were at the point of needing physical assistance from American troops. Not every enemy island was taken in sequence. Some were bypassed, but their supplies were cut off or reduced, a task well suited to PT boats patrolling along island coasts.

Keeping Henderson Field on Guadalcanal operable and building and repairing
roads and improvising bridges was the job of the Sixth Naval Construction
Battalion, or "Seabees." They could repair damage from a 500-pound bomb
in 40 minutes and, if shovels were short, use their helmets to dig.

to attempt one last-ditch effort to give the boats more clout to attack the barge destroyers they were supposed to be sinking, but never could. They would give the PT forces larger-caliber gun nests.

As rumors swirled that the PTs would be given a less glamorous task like shore patrol, Kennedy jumped at the chance to prove they could do more. While *109* was getting its rudder fixed, he asked for a more powerful gun that could give him the killing edge for their next encounter with the enemy. What he got was not nearly so inspiring. In fact, it came from the Army, not the Navy, and it was more suited to a museum than active duty. Kennedy was given a 37-mm breech-loading single-shot weapon designed to hit tanks, not barges. But that was the best that he could do.

Kennedy swallowed hard and tried to find a way to attach this relic to *PT 109*'s deck. What he might have been fearing, but never said to anyone, was that it would take a lot more than even the best guns to give these boats the edge they needed.

The PT Boat Story

Their supporters claim that PT boats were high-speed, versatile, and hard-to-hit assailants, American Davids to the Japanese Goliath, like the English Navy versus the Spanish Armada, invaluable tools in the fight against the Imperial Japanese Navy.

Critics argue that PT boats were unreliable, unarmored, vulnerable, dangerous, ineffective blunderers that did as much friendly-fire damage to Americans as to Japanese.

The truth? With the luxury of hindsight and the examination of many accounts of actions in World War II, the verdict on boats like *PT 109* leans more toward the first viewpoint. Although the PTs were not effective as torpedo boats, they became much more lethal to the Japanese, and in particular their barges, as gunboats.

The fast boats—in design more akin to giant speedboats than to other combat craft—were lightly armed at the war's beginning, with four 21-inch torpedo tubes and four .50-caliber machine guns in twin turrets. Their crew comprised about a dozen men, depending on the particular design. Most of the boats that fought in World War II were manufactured in Bayonne, New Jersey, by the Elco Navy Division of the Electric Boat Company, later part of General Dynamics, although some variations were manufactured by the Higgins Boat Company and Huckin Yacht Company; the latter tended to be shorter, as short as 72 feet. They had marine plywood hulls. The initial World War II PT boat was 77 feet long, the later and more coveted model the 80-footer that Red Fay had risked his naval career to ride at Melville.

Red Fay is a PT defender. "The problem is that they were initially seen as torpedo boats. Because our torpedoes were too slow, that just wasn't working. But as soon as we equipped them with sufficient guns—like when we added twin 50-mm in the turrets and 20-mm in the stern, they could do serious damage against the Japanese barges. Our guns were able to annihilate the barges."

As Curtis Nelson explains in *Hunters in the Shallows,* PTs were born in the American Civil War, evolving out of the spar torpedo boat, a small mechanically driven launch that would fire an explosive charge against the hull of an enemy ship. By the late 1800s, private American boat designers as well as Europeans had made many advances, including propeller-driven torpedoes, the internal combustion engine, and the planing hull that would enable the construction of a rudimentary torpedo boat. The U.S. Navy, however, was the last to recognize the value of such a craft, remaining committed to large ships for coastal defense, so an American PT boat precursor incorporating those technologies didn't appear until well into the next century.

In 1937, the Navy realized that the motor torpedo boats other nations—such as the U.K.—did have had proved successful in the test of battle. An experimental U.S. program to design and build them in a bigger way began under the direction of Thomas Edison's son Charles, Assistant Secretary of the Navy.

Edison's program was unusual in that he was asked to move the boat from concept to deployment in a very short time. Although the number of boats produced in the experimental program was not huge, it was sufficient to help turn the spotlight to these fast, risky boats. The first PT boat, *PT 9,* was operational in the summer of 1940. Even that modest deployment set off a firestorm of protest that threatened to derail the entire program, because *PT 9* was designed by British engineer Hubert Scott-Paine and built by Scott-Paine's licensee in the U.S., Elco. American builders fulminated that domestic industry had been slighted in favor of the British.

As production of all kinds of craft moved to higher levels, the tilt to the British on the PT was eventually forgotten. A dozen PTs were at Pearl Harbor when the Japanese hit in December 1941. And as the Japanese invasion force swept across Asia in early 1942, a PT boat commanded by Lt. (jg.) John Bulkeley rescued Gen. Douglas MacArthur and his family from the ruins of Corregidor as it fell to the Japanese in March 1942. In a 36-hour journey to avoid Japanese capture, much of it behind enemy lines, Bulkeley drove the boat 560 miles, broke through the Japanese blockade, and reached the unoccupied island of Mindanao. From there, MacArthur was airlifted to Australia. And so the legend of the plucky PT boat was born.

Crate-top poker on Guadalcanal provided R and R from getting
shot at or strafed, which in autumn 1942 continued to occur as the
Japanese naval forces pressed attacks to dislodge the Marines.

On her way to convoy troop ships to Guadalcanal in September 1942, *Wasp*
had just launched aircraft when struck by torpedoes from a Japanese submarine.
Ablaze and listing, *Wasp* was soon abandoned and scuttled. The loss of a carrier
was a serious blow, but the Marine reinforcements got through safely. The aircraft
carrier became dominant as the most lethal ship in the Pacific and it changed
the nature of naval warfare.

The low-slung boats were between 77 and 80 feet long and weighed between 33 and 43 tons. Early in the war they were armed with four torpedoes, two .50-caliber turret guns, one 20-mm oerlikon cannon, and two 300-pound depth charges.

The real attraction of the PTs was their speed—40 to 45 knots—driven by three aircraft engines that could deliver between 1,200 and 1,500 horsepower each. But the price of such performance was high maintenance. PT engines were temperamental, requiring constant attention and frequent repairs. Too often, particularly early in the war when the logistics train was not yet in place and many green young crews took to the water for the first time, the PTs were not there when needed.

But at the beginning of the war, the image of the PT boat was unassailable. Bulkeley's evacuation of MacArthur from the Philippines was only the first chapter. After that daring mission, his Philippine squadron fought on bravely into the spring, attempting to harass and derail the Japanese war machine, all the while suffering inevitable defeat, loss of all its boats, and withdrawal of surviving crews to safer waters in Australia in April. Yet according to many accounts, the boats managed to sink Japanese ships while operating in interisland waters and without air cover—and this seeming success against conventional odds was noted by many, including Adm. Ernest King.

In the public mind, the fast, maneuverable, and courageous boats were a welcome antidote to the mournful images of huge, cumbersome American battleships caught unawares at anchorage, burning and sinking to the bottom, or much heralded American air forces sitting helpless on the tarmac while the Japanese turned them to cinders, as in the Philippines and Pearl Harbor.

Another contributor to PT lore was film director John Ford, who as a Naval Reserve officer filmed the defense of Midway. He would later go on to direct *They Were Expendable,* starring John Wayne, in 1945, a paean to the heroic PT crews and boats. In real life, the Battle of Midway was the first sign that Japan could be stopped. Considered by historians to be a turning point in the war, good American luck and strategy and Japanese miscalculation and overconfidence sent four critical Japanese aircraft carriers to the bottom in June 1942. The Battle of Midway saved Midway from a Japanese invasion and enabled the U.S. to begin to take the offensive in the Pacific for the first time.

PTs were certainly present at Midway, but they were more spectators than participants. They never had the opportunity to use their torpedoes against a fleet of huge carriers that remained far out to sea and a Japanese invasion force that never struck.

Marines come ashore in Augusta Bay, Bougainville, on November 1, 1943.
They held enough beachhead to contain Japanese forces, which were further
neutralized by Navy and air power. The Japanese grip on the Solomons
loosened, although troops survived in pockets, and naval forces were active.

The Japanese defeat at Midway suggested that the next phase of the war would not be fought in Hawaii, Alaska, and California, but in the southwest Pacific, where Japan was trying to sever the sea links between Australia and the Allies. That pointed more specifically to the Solomon Islands, and a very different kind of topography—reefs, shallow waters, narrow interisland passages—than military planners had envisioned a decade, or even a year, before. As the American Marines were storming ashore at Henderson Field, Guadalcanal, on August 7, 1942, Admiral King was ordering the first shipment of four PTs to be loaded in Panama for shipment to the Solomon Islands. They arrived at Tulaghi under their own power on October 12, 1942, after having been carried on various craft thousands of miles across the Pacific. A second group of four PTs arrived on October 25.

King sensed that by the fall of 1942 the Japanese advance across the Pacific had been stopped—at Midway, in the Coral Sea, and in New Guinea and the Solomons. But how could he keep up the pressure while the dozens of American aircraft carriers and destroyers on order were still in production and months, or years, away from battle readiness?

King had seen the PT action reports from the Philippines. The Solomons, though more primitive and remote than the Philippines, had some of the same characteristics as a battle theater. The waters there were protected by the two rows of islands that defined the Slot, so naval operations were not deep-sea in nature. Destroyers would have trouble maneuvering in tight interisland passages. The islands were riddled with reefs, which could impair larger deep-draft ships. And just as important, American air cover in the emerging Solomons theater had not yet been put in place. It seemed the PTs could handle this environment better than larger, slower, more exposed warships.

As King was taking these small steps, General MacArthur and John Bulkeley, his PT rescuer, were agitating for bigger things. MacArthur, according to Nelson, was claiming in late 1942 that with several hundred PT boats in the Philippines, he could create a de facto naval blockade that would strangle the Japanese occupation. Bulkeley, meanwhile, was being recognized with a Medal of Honor and had a meeting with President Roosevelt in the fall of 1942, while he barnstormed the country recruiting potential PT crews.

It's not clear how much, if any, impact MacArthur's bold advocacy on behalf of PTs had on U.S. naval plans, because he was an Army man and the two services were not terribly friendly with one another. But he and Bulkeley did have an impact on the popular mind. Nigel Hamilton, author of *JFK: Reckless Youth,* quotes Alvin Cluster, one of John Kennedy's PT squadron commanders in the Solomons, as saying the sales pitch on the power of the PTs in 1942 was far outrunning reality:

"The big thing was MacArthur. If MacArthur had traveled out of the Philippines by any other method, you probably never would have heard of John Bulkeley. And that would have been a blessing. America desperately needed heroes after Pearl Harbor, and they would seize on any exploit or any battle to show how great we were. The only reason PT boats ever got the attention they did was that we had nothing else! They really didn't do a lot of damage. But Roosevelt had to point to somebody, and that's why Bulkeley and PT boats got all that attention."

Just how did the PT boats perform in the early Solomons campaign? It depends on whom you believe. According to Nelson, Bulkeley and his partisans claim that in the Solomons the PTs sank or damaged 1 heavy cruiser, 2 light cruisers, 19 destroyers, and 1 submarine; Nelson counters that a reexamination of records confirms only 1 Japanese sub and 1 destroyer sunk, and 1 destroyer damaged, out of 111 torpedoes fired.

Nelson also points out that the early (pre-August 1943) PT action reports in the Solomons also have a familiar "sameness"—PT boats rush into the action, fire torpedoes, see flashes, then lay smoke and retreat to base. He takes the fact that the Tokyo Express kept on returning to the same areas where PTs had harassed them to mean that the PTs were not a serious deterrent.

But Nelson acknowledges several occasions when they did have an impact in the Solomons. One was on October 13-14, 1942, when a surprise PT attack on a Japanese force including battleships *Kongo* and *Haruna* may have unnerved the Japanese and helped stop a devastating bombardment of Henderson Field. Another success may have occurred on November 14, 1942, during the naval battle of Guadalcanal, when PTs apparently spooked the Japanese and contributed to their withdrawal up the Slot. Another was in early December 1942, when PTs sank the Japanese sub *I-3*, and two days later, when they sank the Japanese destroyer *Terutsuki*.

Most historians now agree that the unimpressive early record of the PT boats in the Solomons was followed by greater success in a new role: that of gunboat. And subsequent theaters like the New Guinea and Philippines campaigns gave the PTs another chance to prove their usefulness.

Most significant, the PTs were already demonstrating their potential as gunboats against Japanese barges in New Guinea by late 1942, and this trend continued on into 1943. The slow-moving, wooden-hulled barges, built by forced labor in Japan's conquered territories, gradually became a lifeline for the overextended Japanese war machine. Because the barges were built to hug the coasts and venture in shallow reef-studded waters, which put them off-limits to big American deep-sea warships, this

Combat artist in the Pacific Lieutenant William F. Draper portrays a landing
at Bougainville. The original caption in the April 1944 NATIONAL GEOGRAPHIC
read: "First Wave! A Jap[anese] Mortar Shell Blasts a Landing Boat. Casualties
Are Heavy, But the Fighting Marines Gain the Beachhead."

Joining the campaign for New Georgia, *PT 109* was ordered to the base on tiny Lumbari Island between Rendova and Munda. It was called "Todd City" for the first PT sailor killed at Rendova. Shelter was tents. Meals were mostly C and K rations. Japanese forces were close; patrols were no longer routine.

move could have thrown off the American plans to stop the Tokyo Express. But the PTs could come in as close to shore as the barges.

Not that the PTs were home free with their targets. Increasingly, the barges were armored and more heavily armed, and could put up a valiant fight when needed. But the odds in a barge fight were more favorable than PTs going toe to toe with Japanese destroyers. The PT action record in the Solomons improved after their conversion to torpedo gunboats, with 146 barges sunk or probably sunk, and more than 80 damaged between July 1943, when the conversion to gunboats began, and October 1944.

The PT kill record in the New Guinea campaign was even more impressive. According to Nelson, PT boats sank or destroyed 486 barges over a two-year period. That huge loss meant that thousands of Japanese troops onshore were neutralized or resigned to starvation in the jungles of the huge tropical island.

PTs also provided important support to the U.S. Army and Navy, as they began their "leapfrogging" campaign up the Solomons and across the north coast of New Guinea toward the Philippines. Leapfrogging, begun in summer 1943, was an important tactical innovation; it meant simply that the U.S. did not need to take every Japanese-held island in the South Pacific in order to control it. By leapfrogging, the "skipped" island in effect became cut off, meaning the U.S. could advance twice as fast and at half the cost.

Using this method, New Georgia was invaded in July 1943, while Kolombangara was bypassed, and Vella Lavella was taken in September. Choiseul was bypassed, leaving the big American guns to focus on the Japanese stronghold at Bougainville, where the American assault began on November 1, 1943. And as the U.S. forces surged ahead, PTs were usually at the forefront.

Once Bougainville was subdued, leapfrogging continued on across New Guinea toward the Philippines, and the decision was made to leapfrog the huge Japanese base at Rabaul, leaving its 100,000 well-armed Japanese troops unengaged, while U.S. air and sea bombardments reduced the once mighty harbor and airfields to rubble. During this unstoppable American advance in late 1943-44, the PTs played an important role. Failed resupply meant an ever weaker and demoralized Japanese enemy.

Having begun his PT career in the swashbuckling early era of torpedo PTs in 1942, John Kennedy was present when the new, more successful gunboat tactics first were put into effect. That was why he made the attempt on the fateful night of his collision to lash a 37-mm gun to his bow. That antitank gun, while not a deterrent to *Amagiri*, could have been useful against the Japanese resupply barges.

The Solomons Campaign

While an officer at young Kennedy's vantage point might have found the fog of battle to be confusing, there was a strategic reason that Japan and the U.S. were fighting in one of the most backward, isolated parts of the world. In the late 1930s, U.S. military planners had correctly concluded that the two nations would go to war one day. They were wrong, however, about where it would take place. None of the U.S. high command had been expecting the Japanese assault that steamrollered over Southeast Asia in a matter of months; the assumption had always been that the big clash would come in the Philippines, where the U.S. had a major military presence.

But at the end of 1941 and early 1942, all that thinking was overturned. The attack of Pearl Harbor temporarily decimated the U.S. Pacific Fleet. The Philippines were surrendered in May 1942 as General MacArthur and his family fled Corregidor. The great American base in the Philippines was lost, and the war had to be fought elsewhere.

So how and where to fight back? The Japanese, with their assault on Midway in June 1942, had sought to make it the mid-Pacific anchor of their outer line of defense reaching from the Aleutians in Alaska across Midway and down south through Micronesia. The Solomons would be the southern anchor of the line, as well as a means to cut off New Guinea, Australia, and New Zealand from the Allies.

The Japanese had established a huge base at Rabaul at the top of the Slot that ran from the Bismarck Archipelago northeast of New Guinea to about 500 miles west to Guadalcanal. But they needed to push farther east to be totally secure. In early May 1942, they moved to establish a new base at Tulaghi, a few miles north of Guadalcanal on the other side of Ironbottom Sound.

This strategy, they thought, would anchor the eastern Solomons in their defensive line, and protect the northern coast of New Guinea. From New Guinea they intended to stage an overland march on Port Moresby. The Tulaghi move generated an American countermove: an aerial bombing of the new base from the U.S. carrier *Yorktown*. And so began the first great carrier battle of World War II, the Battle of the Coral Sea.

The battle was a series of clashes that began May 4 and reached a climax four days later. The big stars were carriers such as *Yorktown, Lexington, Zuikaku, Shokaku,* and *Shoho.* Although it would redefine the concept of naval warfare and relegate battleships to a secondary role, this great battle was not decisive but more a standoff of two combatants who had a hard time even finding each other in the rainshowers and vastness of the western Pacific. At the end, the Japanese light carrier *Shoho* and the U.S. carrier *Lexington* sank to the bottom of the sea. In tonnage, the

U.S. losses were greater. Yet, unaccustomed to defeat or stalemate, the Japanese felt their loss more deeply, and because the Americans had held their ground and the Japanese held off their invasion of New Guinea, the battle was spun as an American victory. Later, as the pattern of American victories became more defined, Coral Sea would be seen as the place where the Americans had finally drawn a line that the Japanese could not cross.

This May 1942 encounter was followed by America's first big amphibious operation in the Pacific, to be followed by dozens more. The August Marine invasion of Tulaghi turned that island into an American base, where Jack Kennedy undertook his Solomons training a few months later. The same military operation seized the Japanese airfield on Guadalcanal and renamed it Henderson Field, later the site of the Solomons' capital, Honiara.

But taking the airfield and renaming it did not mean that Guadalcanal was won. In fact, the Japanese, rather than surrendering the island, regrouped on either side of Henderson Field and staged a series of bloody, often suicidal assaults to try to recover it. Both countries sent in thousands of reinforcements to stake their claim to the malaria-ridden beaches and jungles of Guadalcanal.

One bloody Japanese incursion to the east of Henderson Field was undertaken by a group originally trained to invade and hold Midway. More than 900 crack fighters charged ashore into a hail of American lead. About 800 Japanese died, while only a few dozen American Marines were lost. The next big battle was jungle fighting at the Tenaru River in August, and again the Americans won. For the 16,000 Marines fighting the heat and deprivation of the tropics, this victory was especially rewarding. They had proved that the Japanese jungle fighting skills, elevated to legendary status by victory after victory in the Philippines, were not invincible.

Yet the awful slaughter on Guadalcanal continued. Japanese troops were prepared to die rather than surrender, but this dedication was not backed up by the best Japanese strategy. Again and again the Japanese high command would squander an advantage. One example came after the American naval debacle at the Battle of Savo Island the second week of August 1942. Japan could probably have retaken Henderson Field, because the U.S., at least temporarily, had no Navy or Air Force to hold them off. But instead, the Japanese steamed back to the safety of Rabaul, providing the Americans with an opportunity to regroup.

By the time John Kennedy arrived, these Japanese strategic mistakes had allowed the Americans to push their foe several hundred miles back toward Rabaul, to Gizo

"The dense jungles of New Georgia made for miserable fighting conditions. The soldiers were wet from the constant rains, slept in the mud, and casualties were heavy from both combat and malaria," wrote a historian of the battles there.

Marine and Army troops seized the island of Rendova in the New Georgia campaign. The action was fierce but swift, and troops in July 1943 had time to eat hot meals before the next invasion. PT boats were assigned to disrupt supplies bound for a Japanese airstrip on nearby Kolombangara Island.

and New Georgia and Kolombangara, where *PT 109* would undertake her final duty. But Japanese mistakes weren't the only cause of their retreat.

Americans had achieved success at code breaking. As a result, Japan's most secret communications were being read by Americans. This newfound intelligence resulted in John Kennedy's orders the night of August 1—to intercept and harass the Tokyo Express.

Adding eyes and ears to radio code breaking were some little-known heroes. Known as the Coastwatchers, they were a group of heroic Australians who were scattered at critical intervals on the shores and jungle hillsides of the Solomons and New Guinea, risking their lives to spy on the enemy.

The Coastwatchers had originally been conceived in the 1930s by the Australians, when they realized that whoever controlled the Solomons and New Guinea would be able to control, and even cut off, shipping to Australia's east coast ports. So an effort was begun to construct a line of volunteer observers strung out over a distance of nearly 2,600 miles, equipped with 300-pound pre-transistor "teleradios," generators, and supplies.

The bulky generator-powered teleradios were the mainstay of communications for the Coastwatchers, although they also depended on native couriers when necessary. Trying to station the Coastwatchers and get their radios in place without succumbing to salt water, jungle climate, equipment breakdown, or Japanese surveillance was an unending challenge, and one that was met surprisingly well. The genius behind this strategy was Lt. Comdr. Eric Feldt of the Royal Australian Navy.

The network was in place when war broke out, but no one anticipated that the islands would be overrun so quickly by the Japanese. With Rabaul becoming a Japanese stronghold in early 1942, Japanese ships and troops quickly pushed toward Guadalcanal at the far end of the Solomons.

The Coastwatchers suddenly found themselves reporting from behind enemy lines, particularly on Guadalcanal and Tulaghi in spring 1942, as thousands of Japanese troops and military construction units swarmed ashore to build a Japanese airfield. Most of the Coastwatchers on Guadalcanal were able to withdraw deeper into the island's interior as the Japanese advanced, relying on sympathetic islanders to report back on the kinds of ships and construction equipment coming ashore. Yet as the Japanese became more entrenched in the summer of 1942 and the Allies seemed powerless to stop them, islander loyalties began to waver.

Through informers, and using threats and coercion, the Japanese learned of the existence of the Coastwatchers, and even learned many of their names. Using native go-betweens, they sent warnings that they would soon pursue the Australians and that their days were numbered. According to Walter Lord's *Lonely Vigil*, one Coastwatcher learned that a meeting of village chiefs had been held and the majority wanted to turn in the Australian to protect their own villages from retribution. But a single chief had persuaded them not to do that—not out of any great loyalty to the Australians, but by warning the chiefs that the well-armed Australian was a good shot and would kill too many of them during the capture to make it worth attempting.

At the war's outset the islanders' loyalty to the British and their Australian allies wasn't strong, except among those who had benefited from colonial rule. No one knew whether the rapacious Japanese would fare any better in winning the hearts and minds of the population. Still, the Japanese dragooned able-bodied men into helping build the airfield without paying them for their labors, and that didn't earn them any respect or loyalty among the islanders.

With the situation deteriorating in July 1942—the Japanese coming closer and the islanders becoming less dependable, if only because they had to placate their new Japanese occupiers —most of the Guadalcanal Coastwatchers requested permission to evacuate their positions. Instead, their commanders urged them to hang on just a few more weeks, hinting that a "surprise" was coming that would save them from Japanese capture. And in that interval, the watchers were able to transmit valuable intelligence to the Allied command about the state of the Japanese airfield and what kinds of weapons were protecting it. The native helpers, many of whom had never seen even a bicycle, became adept at describing exotic construction vehicles, guns, and aircraft.

The U.S. invasion of Guadalcanal began early on the morning of August 7. Almost 17,000 Marines stormed ashore, seizing the new Japanese airfield and beginning the turn of the tide on the island. Although the Guadalcanal airfield was taken fairly easily, the fight for Tulaghi took a bit longer. The Japanese dug in, using caves and bunkers, giving a taste of the kind of fighting that would come later in places like Tarawa and Iwo Jima.

Despite this small but important victory in the east, the Australian Coastwatchers' position remained perilous as the pulse of Japanese expansion and the Tokyo Express continued. On the morning of August 8, watchers on Bougainville, several hundred miles to the west of Henderson Field, saw the Japanese counterattack coming in the form of a mass of Japanese Zeros flying south. They warned the Allies via their trademark teleradios and gave them an hour and a half to get ready, though the defenders were still far from prepared for what came. The Guadalcanal Coastwatchers were also eyewitnesses to the terrible naval battle that raged near their island on August 9, 1942, when Admiral Mikawa and the Imperial Japanese Navy sent the *Canberra, Astoria, Vincennes,* and *Quincy* to the bottom of Ironbottom Sound between Guadalcanal and Savo.

On the morning of August 14, one of the Guadalcanal watchers, British colonial official Martin Clemens, responded to an Australian order to come out of hiding and seek refuge with the Americans. He marched a small group of his loyal islanders right up to the American lines. The shell-shocked Marines, desperately trying to hold their Guadalcanal beachhead and not accustomed to friendly overtures, were at first uncertain what to do with the visitors. Clemens and his group were finally allowed within the lines. According to Lord, Clemens's elation at joining his fellow Allies and finding what he thought was a refuge turned into despair when he realized how vulnerable the Marine beachhead was. In fact, even with 17,000 men, the American command felt it was only able to hold the airfield and a small perimeter around it. The Japanese still controlled much of the rest of the island and the air above and the oceans around. Even worse, supplies and reinforcements were not forthcoming. In fact, the American outpost looked rather desperate and miserable from the inside.

As the days wore on, the Japanese poured men into positions to the east and west of the beachhead and Henderson Field. The objective was to surround the Americans and eliminate them. But the Guadalcanal Coastwatchers and their native helpers, including islander hero Jacob Vouza, who survived an inhuman bout of stabbing and torture at the hands of the Japanese, were able to gather enough intelligence about the Japanese moves to keep the Americans prepared and alert. No Japanese attempt to overrun the American base succeeded, and most were terrible slaughters.

Another unofficial Coastwatcher was Catholic priest Emery De Klerk on the far west end of Guadalcanal. Ordered by his superiors to remain neutral in the war, he concluded the Japanese occupiers would be worse for his congregation than the British Empire, so he tilted more and more toward the Allies, providing valuable information, support, and, through his good ties to the islanders, manpower.

Looking at the changed map in the fall of 1942, Coastwatcher commander Feldt and his fellow officers became concerned that the Coastwatchers were leaving too many miles unobserved from Bougainville to Guadalcanal. With the Japanese struggling to build bases in that stretch to better support their Tokyo Express resupply, the danger was that operations would be mounted that Coastwatchers couldn't see. So they decided to put a man on each of the intermediate islands of Vella Lavella and Choiseul to the north. The trick was how to get them into those islands that were still in Japanese territory.

Rear Adm. John McCain, father of the future U.S. Senator John McCain, ruled out using aircraft. An American submarine, *Grampus,* was selected instead for the duty. On October 14, it unloaded two men, 2,000 pounds of equipment, and several small boats under cover of night. Then the sub slipped away. (Sunk in March 1943, it is believed to rest on the seafloor not far from where *PT 109* went down.)

The two Coastwatchers were in for an adventure. As the hours of night wore on toward morning, they couldn't find a good landing place. Knowing that daybreak would probably bring Japanese detection, they were forced to shoot the breakers over a coral reef, losing one of their three boats in the process. The radio arrived safely, but the men floundered in the water, saving as many of the other supplies as they could. They hid everything in the shore foliage just before a Japanese patrol plane flew by on its morning rounds.

A week later, they had set up shop at an abandoned copra plantation, with the radio set to transmit from atop a 3,000-foot (915-meter) peak. Just when everything seemed set, the radio wouldn't work, rendering their presence temporarily useless.

The two Coastwatchers on Choiseul had similar landing problems and barely escaped detection by the Japanese. By a stroke of good luck, one of the Coastwatchers met a former native associate, and with at least that foothold, they began to wait for the right time to move their equipment farther inland.

In November, Coastwatchers on Bougainville watched helplessly as the Japanese assembled one of the greatest operations of the war, a final desperate and massive attempt to destroy the Americans at Henderson Field. Coastwatchers reported seeing 61 ships, including 33 destroyers, huge transports that carried 12,000 men, tenders, and others. As a result of steady reinforcements in late October, Coastwatcher warnings, and American code breaking, Americans were able to repel the northern part of the Japanese operation in the vicious naval battle of Guadalcanal in early November. The huge Japanese invasion fleet from Rabaul and Bougainville was bombed continuously as it moved east and south. The bombing took a heavy toll, sending half the Japanese transports to the bottom with their men. A few made it to Guadalcanal, but by then the invasion of 12,000 men had melted to just 2,000—not enough to make much of a difference. Even more ominous for the Japanese survivors, hardly any food supplies made it through the American meat grinder, meaning the increased numbers of men would be even hungrier and more beleaguered.

By late 1942, the 25,000 Japanese on Guadalcanal were foraging for food, so successful was the American choke hold on the Tokyo Express. Increasingly, the Japanese looked to build intermediate air bases between Rabaul and Guadalcanal in

B-17 Flying Fortresses, so critical in Europe, had a limited role bombing enemy onshore facilities and scouting for enemy vessels. High-altitude bombing of ships was difficult. When one Japanese destroyer was struck and sunk, its commander remarked, "Even the B-17s could make a hit once in a while."

order to increase the flow of supplies into their dying eastern stronghold. They would make an attempt at Munda on New Georgia Island, halfway between Rabaul and Guadalcanal. But the Allies sent in more Coastwatchers to keep an eye on their movements and thwart the attempt in any way they could.

By the time John Kennedy arrived in the Solomons in March 1943, the Japanese had quietly surrendered Guadalcanal, sneaking their starving and demoralized men out under cover of night. But just because the front had shifted northwest by a few hundred miles didn't mean the fighting was any less intense or the Tokyo Express any less dangerous. Some officers were heard to quip that if the Allies kept up the current pace of advance, it would take another ten years to get to Japan.

To turn the tide more quickly, more Coastwatchers were implanted in the critical mid-Solomons gap by mid-1943. One was Sub Lt. Arthur Reginald Evans of the Royal Australian Navy, based on the hillsides of volcanic Kolombangara, overlooking the critical Blackett Strait, where the Tokyo Express was known to rove, and only a few miles from Lumbari Island, where John Kennedy was stationed. Evans would play a critical role in Kennedy's rescue, because he would see the brief flame from the collision's explosion, even though he didn't know what it was at that moment.

The creator of the increasingly desperate marvel of naval logistics called the Tokyo Express was Rear Adm. Raizo Tanaka. He used almost anything that would float—destroyers, cruisers, submarines, and barges—to get the supplies to the Solomons front. But the challenge of feeding, clothing, and arming hundreds of thousands of Japanese soldiers and sailors scattered up and down the Slot, on New Guinea, and in the Philippines, as the Allies pounded away relentlessly, was eventually beyond his ability. It was not possible, even had the Japanese cut down the rations and supplies to a bare survival level. By November 1942, he was already incapable of meeting the needs on Guadalcanal alone. The Japanese 17th Army's staff calculated its supply needs as being 5 destroyer loads per night, or 150 loads per month. This, of course, depended on air cover to protect the ships, and this was not available. Although freighters were be capable of carrying more, they were more vulnerable to attack. In the end, the Japanese logistics challenge in the Slot became overwhelming. But they did try, through 1942 and most of 1943, and the reefs, channels, and jungles of the Solomons and Papua New Guinea are littered with the rusting detritus of the Japanese attempt. As the moments ticked down to the long night of August 1-2, the Tokyo Express, though ultimately doomed, was still a force to be feared.

THE

CHAPTER THREE

COLLISION

Skipper Kennedy relaxes in the cockpit of *PT 109,* which had no canopy, no armor, no radar, and only basic instruments. Off duty, he had driven his mates nuts by playing the record "Blue Skies" over and over, but on August 1, 1943, *PT 109* entered Blackett Strait under the cover of darkness.

AUGUST 1-2, 1943- NOVEMBER, 1963

AUGUST 1, 1943, WAS NOT MUCH DIFFERENT from the many hot tropical days John Kennedy and his men had already experienced in the wartime Solomons. The news of another big Tokyo Express run, gleaned from American code breaking, had come in. The afternoon was a frenzy of preparation for *PT 109*, in particular the addition of the 37-mm Army gun. Kennedy didn't have time to get it bolted down, thanks to a Japanese air raid that sent men diving for dugouts and cover or manning guns to answer the fire. Japanese dive-bombers roared perilously nearby, several going down in spectacular crashes while bombs and automatic weapons rattled the air. Kennedy and the undamaged boats were forced to pull out of the base for a few minutes until the all-clear sign was sounded. Once the raid was over, the crew only had time to lash down the new gun on the bow with line, giving up the life raft that was normally stowed there.

Later Kennedy and the other PT skippers went ashore for their afternoon briefing with Warfield. Because of the ominous warnings from his commanders, Warfield was deploying all fifteen available PTs; *109* was not originally scheduled to go out that evening. In part this was done to counter the size of the Japanese flotilla, but he was also responding to his commanders' belief that the Japanese would be gunning for the PTs that night, and he wanted an extra complement to keep the enemy busy.

Returning to the boat, Kennedy made an impromptu change to *PT 109*'s crew, which had been changing its roster from week to week, the result of reassignments, rotations, and other routine personnel shifts. Barney Ross, formerly executive officer on Bill Battle's *PT 166*, which had gone down only a few days before, asked Kennedy if he could ride along that night. The skipper agreed.

At 6:30 that evening when *109* and the other boats pulled away from their docks, 12 men were under Kennedy's command. Because so many reassignments had occurred and no final crew list was submitted to Warfield's office on August 1, the names had to be assembled after the fact. They included: executive officer Ens. Leonard "Lennie" Thom, from Sandusky, Ohio; Ens. George "Barney" Ross, lookout and 37-mm gunner; machinist's mate Patrick McMahon of Los Angeles; gunner's mate Charles Harris of Boston; machinist's mate Gerard E. Zinser of Belleville, Illinois; radioman John Maguire of Hastings-on-Hudson, New York; machinist's mate William Johnston of Dorchester, Massachusetts; ordnanceman Edgar Mauer of St. Louis; torpedoman Ray L. Starkey of Garden Grove, California; Seaman 1st Class Raymond Albert of Cleveland; motor machinist's mate Harold Marney of Springfield, Massachusetts; and Andrew Jackson Kirksey of Macon, Georgia.

In one of the most fateful decisions of the night, according to Red Fay, Kennedy made Barney Ross the forward lookout as well as the gunner for the 37-mm lashed to the deck. Kennedy didn't know the man was night blind.

Dick Keresey was in the same operation, the Battle of Blackett Strait. He was only a few miles from *PT 109* when Kennedy was hit and sunk, and he devotes a chapter of his book *PT 105* to how things looked from his perspective.

The account that offers the closest rendering of what Kennedy and his men experienced during those critical days and nights is Robert Donovan's exhaustive *PT 109: John F. Kennedy in World War II,* published in 1961. Although later, revisionist accounts have called into question some of Kennedy's actions and decisions before and after the collision, those accounts did not have the cooperation of Kennedy or any of his crew. According to Donovan, his upcoming fate was heavy on the mind of crewman Andrew Kirksey that night. The Japanese attack of July 19 in Vella Gulf had spooked him, and he had been telling his mates for several days that he didn't think he'd live much longer. As they steamed away from the docks into the darknening night, Kirksey was even more convinced of his approaching end. When others suggested he stay onshore, he refused for fear of being called "yellow."

Fifteen PT boats departed from Lumbari that night. Kennedy was assigned to PT Division B, under the command of Lt. Henry Brantingham on *PT 159.* Brantingham was a veteran of the first PT service in the Philippines, and had taken

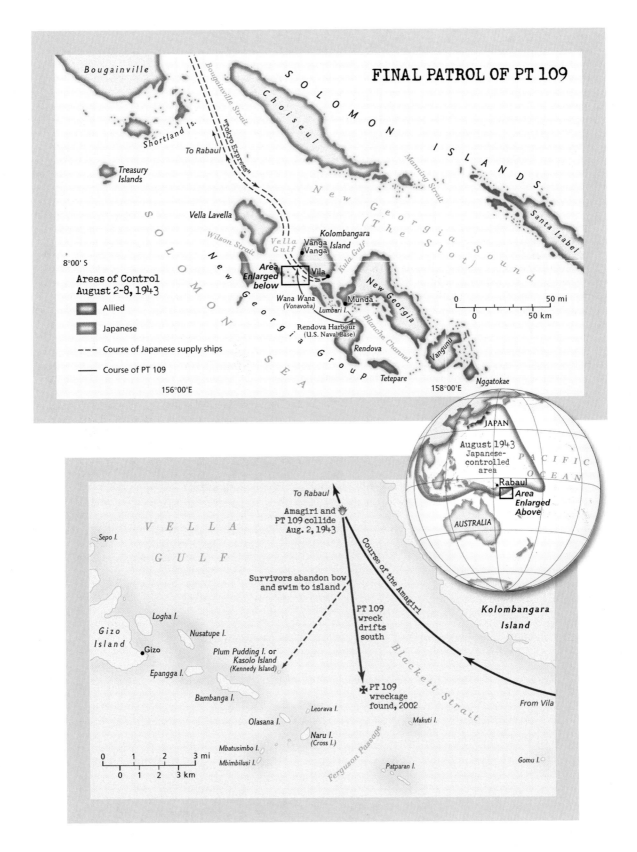

FINAL PATROL OF PT 109

Bougainville

SOLOMON

Choiseul

ISLANDS

Shortland Is.

Bougainville Strait

"Tokyo Express"

To Rabaul

Treasury
Islands

SOLOMON

Manning Strait

Santa Isabel

New Georgia Sound (The Slot)

Vella Lavella

Wilson Strait

Vella Gulf

Kolombangara
Island

Vanga
Vanga

Kula Gulf

8°00' S

Area
Enlarged
below

Vila

Areas of Control
August 2–8, 1943

Allied

Japanese

Course of Japanese supply ships

Course of PT 109

Wana Wana
(Vonavona)

Lumbari I.

Munda

New
Georgia

Rendova Harbour
(U.S. Naval Base)

Blanche Channel

Rendova

Vangunu

0   50 mi

0   50 km

156°00'E

SOLOMON SEA

NEW GEORGIA GROUP

Tetepare

158°00'E

Nggatokae

JAPAN

August 1943
Japanese-
controlled
area

PACIFIC

OCEAN

Rabaul

Area
Enlarged
Above

AUSTRALIA

VELLA

GULF

Sepo I.

To Rabaul

Amagiri and
PT 109 collide
Aug. 2, 1943

Course of the Amagiri

Kolombangara
Island

Logha I.

Gizo
Island

Nusatupe I.

Survivors abandon bow
and swim to island

PT 109
wreck
drifts
south

Gizo

Plum Pudding I. or
Kasolo Island
(Kennedy Island)

Blackett Strait

Epangga I.

Bambanga I.

Leorava I.

PT 109
wreckage
found, 2002

From Vila

Makuti I.

Olasana I.

Naru I.
(Cross I.)

0   1   2   3 mi

Mbatusimbo I.

Ferguson Passage

0   1   2   3 km

Mbimbilusi I.

Patparan I.

Gomu I.

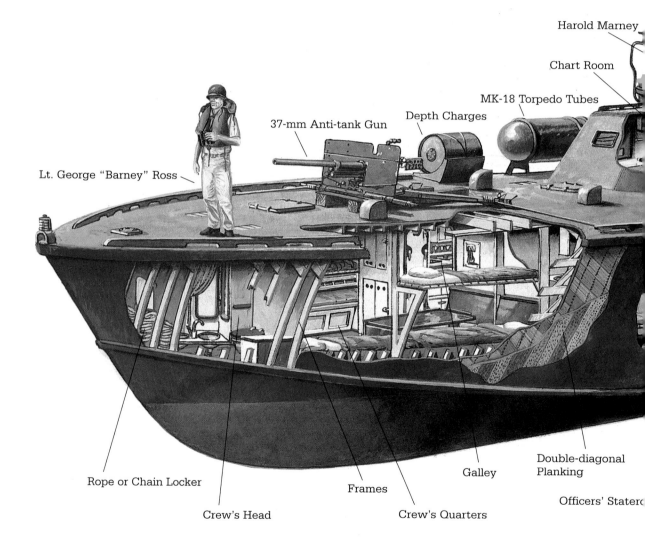

Harold Marney

Chart Room

MK-18 Torpedo Tubes

37-mm Anti-tank Gun

Depth Charges

Lt. George "Barney" Ross

Rope or Chain Locker

Crew's Head

Frames

Crew's Quarters

Galley

Double-diagonal
Planking

Officers' Stateroo

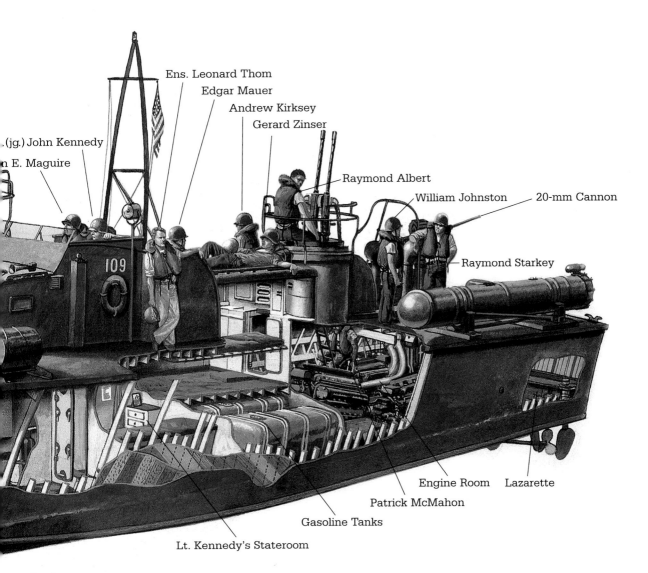

Ens. Leonard Thom

Edgar Mauer

Andrew Kirksey

Gerard Zinser

(jg.) John Kennedy

n E. Maguire

Raymond Albert

William Johnston

20-mm Cannon

Raymond Starkey

109

Engine Room

Lazarette

Patrick McMahon

Gasoline Tanks

Lt. Kennedy's Stateroom

ers' Head

part in evacuating General MacArthur and his family from Manila. The other two boats in Division B were *PT 157* and *PT 162.*

As was standard, only Brantingham's boat had radar; the other three in his group were dependent on radio communications and visual or even shouted signals from the lead boat. The last glow of day illuminated the silhouettes of Rendova and Kolombangara, but a low shelf of cloud quickly overran the area over Blackett Strait, lair of the Tokyo Express, killing all sky glow. The decreased visibility made it that much more difficult for the boats to remain oriented. No surprise there, however; the PTs had grown used to undertaking their missions, and sometimes getting lost, in the pitch black of the overcast tropical night.

By midnight, the men on *PT 109* and *PT 162,* commanded by John Lowery, felt as though they were isolated from everything else on Earth. Radios transmitted only static; the crew couldn't see a thing out on the treacherous Blackett Strait. After a time on the water, they were not even sure where their leader, Brantingham, was, and so they waited, talking politics. Radioman Maguire stayed close to Kennedy, as though he had something to report, but he did not.

Kennedy and the other lost boats waited where Blackett Strait was its narrowest, a five-mile funnel between Kolombangara on the northeast and Gizo and the little islands that had no agreed-upon names to the southwest side. The logical assumption was that if the Tokyo Express was to be spotted, it would be here.

Dick Keresey was well aware of the problems of seeing anything in the dark Blackett Strait. "Later on we learned to operate closer in to shore, so that we could look back at the horizon out to sea and have better visibility. Looking north toward Kolombangara, you could only see the black bulk of the island. You couldn't see any ship moving against it. But if the Japs were positioned closer to shore than we were, they could see us against the southern horizon."

While the others floated in darkness, Brantingham, thanks to his radar, knew that danger was approaching. In the early hours of August 2, his primitive radar, designed for aircraft use, detected four shapes moving up the strait from Vila Point, near the invisible volcanic massif of Kolombangara.

Brantingham saw the images but did not know what they were. All he knew was that something was out there. Moving fast.

Guessing from their apparent small size and position close to shore that they were Japanese barges, he set off in pursuit, armed with Mark VIII torpedoes, relics from World War I. He fired the torpedoes, the best he could throw at them. Far from hitting the enemy, one of them set Brantingham's own torpedo tube afire. The firelight thrown on his boat drew return fire from the mysterious four shapes on the black horizon, forcing Brantingham to take evasive maneuvers and lay down smoke barriers to keep from being killed.

Various accounts confirm that Kennedy in *PT 109* and Lowery in *PT 162* had lost contact with the commander of Division B. *PT 169*—under the command of Philip Potter, from Division A—became lost as a result of poor communications and wandered into the area.

Before they departed, they had planned that Brantingham would radio Lowery with instructions on how to respond to any developments, and Lowery would pass the information on to Kennedy. It was around two in the morning, and if Brantingham had sent out a radio message to Lowery about what he'd seen or what he was doing, *PT 109* never received one. Dick Keresey says that no such message was transmitted, because if it had been, he would have heard it on *PT 105*. As their squadron commander fought his own engagement, *PT 109* continued to drift, silently.

The three lost boats did see intermittent flashes off on the horizon. On the radio they heard unattributed snatches of conversation about enemy contact and firing torpedoes. Keresey, who was some miles away, remembers those radio reports as completely garbled. "We were under radio silence," he recalls, "but then these people come on the radio and start shouting about how terrible it is, they are under attack, they are retreating. It was totally useless information, you couldn't make any sense out of it, who it was or where they were. It was scaring the hell out of me."

Kennedy and crew believed that the flashes were from Japanese shore guns, and so they moved away from Kolombangara. When searchlights from one of the four Japanese destroyers cut the dark nearby and shells exploded, Kennedy and Lowery concluded that the Japanese were also coming within range of the shore guns, exposing themselves to friendly fire.

The barges were, in fact, four of Japan's deadliest destroyers—*Amagiri*, *Hagikaze*, *Arashi*, and *Shigure*. They had unloaded tons of supplies and hundreds of troops at their Vila Plantation base and were making the return run up to the Slot to Rabaul.

The three lost PT boats radioed back to Rendova and asked for new orders. The response was simple: to patrol in Blackett Strait. But exactly where was unclear. Kennedy first accompanied the two boats north to Vanga Vanga, then led them back to the approximate location where they had lost contact with their respective groups. And they waited.

A strange peace had fallen on *PT 109*. Radioman Maguire was quietly saying a rosary. Kennedy was at the wheel. Only one engine was engaged, and the boat moved almost soundlessly through the night at low speed. Barney Ross was manning the lashed bow gun, peering off into the void, plagued by the night blindness he hadn't reported to Kennedy. Patrick McMahon was in the stifling engine room tending the three Packard engines. Harris, Johnston, and Kirksey, off duty, were either asleep or lying on the deck at various points.

According to Donovan, Harold Marney, who was in the forward gun turret, suddenly shouted, "Ship at two o'clock!" Kennedy, looking to his starboard, could see a black shape approaching. He concluded that it was one of his fellow PT boats.

But the thing kept getting bigger, and it was going fast—30 to 40 knots, according to the Japanese. In that sickening interval as the dim shape grew, its bow wave phosphorescent from the tropical marine creatures disturbed within, Kennedy and the others were forced to understand what was about to happen. With only 40 seconds until impact, John Kennedy finally realized that a Japanese destroyer was bearing down on at him at full speed.

Initially, it seemed as though the destroyer would pass on a parallel course. But *Amagiri,* on the orders of her commander, had turned to ram the smaller PT boat. He and the other Japanese commanders had chosen this route because they would rather engage the small and unreliable PTs in Blackett Strait instead of American destroyers, which were probably stationed out in the Kula Gulf. He could not fire a torpedo at this impertinent little boat—the PT was too close and the torpedoes would sail underneath their target—so the Japanese skipper made a torpedo out of his ship.

Ideas of using his own torpedoes passed through Kennedy's mind too, and he gave orders to get the boat in position, but they didn't have time. His only option was to fire a gun. Barney Ross tried to fire, but the breech of the 37-mm was closed and the seconds were whirling away. He too knew that he was out of time.

Watching a collision of this magnitude unfold was terrifying. With 20 seconds left, Kennedy ordered general quarters. Those who heard it tried to do what they had been trained to do. But there was no time.

The huge destroyer towered over its prey. A big ship moving at 30 knots not only moves water, it moves air. The bow wave must have been preceded by a gust befitting the meaning of *amagiri* in English, "divine mist." Those on board believed that the huge prow slammed into *PT 109* on the starboard side, right at the cockpit.

A view taken from *PT 109* shows a crewman, at right, sending hand signals from *PT 61*. Such signals were no use at night when PTs operated. On the night of August 1, 1943, four divisions of PT boats patrolled Blackett Strait and only the four lead boats had radar.

PT 109 was one of nearly 300 80-foot-long patrol torpedo boats built for the U.S. Navy in World War II. They were agile and could sting, so they were nicknamed "Mosquito boats." Constructed much like civilian speedboats and yachts of their day, with wood planking fastened over wood frames, PTs shared nothing of their creature comforts.

Thousands of tons of steel moving at high speed when hitting a nearly motionless and much smaller hull of wood will cause catastrophic damage. The huge screws of *Amagiri* were turning at full force, pushing the prow ahead like a great samurai sword.

The PT boat, hit full force, lost its starboard side. Weighted down by an engine and without any means of maintaining floatation, it sank immediately into the sea. The larger part, sustained by watertight compartments, though knocked violently askew by the collision, stayed afloat, though unable to gain headway or fire.

As for the men on board, Harold Marney, taking his first turn ever in the forward gun turret, was crushed by the Japanese hull and never seen again. Young Andrew Kirksey of Georgia, resting on the starboard side slightly toward the stern, who had sensed his own death coming, had his premonitions come true as the great enemy hull slammed into him. "Pappy" McMahon of Los Angeles, down tending the hot and noisy engines, saw the engine room gape open and a torrent of flaming gasoline surge at him before he was thrust into the cold, salty abyss of the ocean. William Johnston was asleep one second, and the next thrown into the ocean wearing helmet, shoes, and, mercifully, a Mae West life jacket. He wanted to let go and die, but he thought of his wife: "Nat will think I'm yellow." Rather than have his wife think him a coward, he slowly fought his way to the surface. It seemed to take a lifetime, and he was tempted to surrender. But finally he breached the waves, swimming in the wreckage temporarily lit by fire, thinking he was all alone.

Charles Harris, sleeping with his life jacket as pillow, woke from deep sleep just as the ships collided. Seeing the huge prow bearing down, he tried to dive overboard, but his leg was slapped with a disintegrating plank of hull or deck or torpedo tube. His life jacket in hand, he felt his injured leg go numb and then dead. That left him swimming in the dark sea, crippled. He too thought he was alone, until he saw another spectral figure, this one a man roasted by fire and so shrouded by darkness that he could not be recognized.

Gerard Zinser was thrown neatly out onto the black waves, and floated in the ocean alive. Torpedoman Ray Starkey was smacked in the head by something, but stayed on deck, falling perilously near the gasoline-driven fire. He slipped into a stupor. Barney Ross, who had never gotten his chance to fire that infernal 37-mm gun, was hurled about the deck. Mauer and Maguire were flung with him, thrown like dolls. Out in the water, broken by patches of fire, floated others, including a temporarily delirious Lennie Thom.

The skipper himself, young John Kennedy, was thrown backward into a bulkhead in the cockpit, but remained on deck. Seemingly oblivious to the severe blow he'd taken to his back as he was thrown, Kennedy was one of the ablest men on the wreck. He began issuing orders, making decisions, moving about, and helping those he could see and assist.

As historian Richard Frank says of the collision, "They were fantastically lucky that the 100-octane gasoline tanks didn't ignite in a fireball and incinerate all of them. There was a fire, there was injury to at least two members of the crew from fire, people swallowed gasoline or the fumes; it was overpowering. We're talking a second or two, a foot or two between Kennedy being killed and Kennedy surviving, just a tremendously traumatic event."

As the flames blazed all around him, triggering fears of a fuel tank explosion, Kennedy's first order was for the men to get into the water, which from his vantage point initially seemed a reprieve from a flaming, sinking boat. But the water, itself on fire, offered no refuge; fumes filled his lungs and choked him. Rather than swim in an ocean surrounded by flames and fumes, Kennedy instead had Maguire and Mauer climb back on the hull with him. One of them shone a light off into the night, searching for any survivors.

Where were the other PT boats in the group? Surely they had seen the collision; surely they could see flames.

From his vantage point to the southeast, Dick Keresey did see a flickering light on the far water, but he thought he was seeing a defective flare. "I'd never seen as many flares as I did that night," he said, 59 years later. "The Japanese were really active." He watched the light flicker for a few moments and then disappear. He had no idea it was a burning PT boat, and that men were overboard in need of help.

Nor did anyone else. A quick rescue for Kennedy and his men was not to be. *PT109* was lost and in great danger, the dark ocean beneath, the obsidian sky above. As the minutes following the collision crept by, the flames burned lower and less fiercely, and the perimeter of darkness crept in closer.

Kennedy put on a life belt and went back into the water, looking for survivors. Stray voices could be heard in the darkness by those in the water or on the boat. Zinser called out about seeing Lennie Thom. Ross came out of his daze and could see Zinser and Thom. Maguire tied on a line and swam out to the voices.

Harris was floating with his dead leg in the dark, looking at the spectral face that drifted toward him until it was close enough to touch, a scorched man in a helmet. It was Pappy McMahon. Because the older man was too burned to do it himself, Harris pulled Pappy's helmet off. But for his life jacket, McMahon would have gone down with the starboard side.

Captain Yamashiro (above), commander of the 11th Destroyer Flotilla, was senior officer on board the *Amagiri* the night she rammed *PT 109*. He shared his recollections in 1961 with Robert J. Donovan and Elliot Erwitt; here he reenacts the crash with hand motions. The Japanese destroyer was photographed at sea in the 1930s (below).

Previous pages: The Japanese destroyer *Amagiri* appears out of the black so suddenly that she cannot bring weapons to bear on *PT 109*. In a report of the action, Lt. B. R. White wrote " ... when *PT 109* had scarcely turned 30 [degrees], the destroyer rammed the PT .... "

The fire died out. Everyone was lost in darkness. Harris could hear other voices and he called out for Kennedy, who used the sound of Harris's shouts to orient himself. When Kennedy got to him, Pappy was ready to surrender to death. He told Kennedy to go on without him. But the skipper pulled the older burned man by his kapok, and used the voices of the men on the floating hull to guide him back. Harris dropped behind, burdened by his dead leg and heavy clothes. He too was about to surrender to the darkness.

"Where are you, Harris?" Donovan quotes Kennedy as calling. "I can't go any further," Harris replied. To which Kennedy answered, "For a guy from Boston, you're certainly putting up a great exhibition out here." That infuriated Harris. Kennedy came back and helped him strip off the extra clothes.

Lennie Thom brought in William Johnston, who had made it up from the suction of the sinking stern to the surface, only to swallow and inhale gasoline. Thom brought him back very slowly, not a good swimmer himself. Ray Starkey was the last one to come in, under his own steam though burned on the face and hands. He had found a floating mattress first, and then spotted the floating hull. He swam what looked to him like several hundred yards.

The 11 survivors were now all together. Kirksey and Marney were gone forever. Johnston was retching himself inside out from gasoline, Pappy was burned over face and hands and arms, shivering with a chill. The list of the bow was increasing. Daybreak was coming upon them.

Being exposed to whatever came flying or sailing by was another fear. Though the Americans had bases and occasional ships in the area, the big islands of Kolombangara and New Georgia were held by the Japanese. Lumbari was miles away.

On the morning of August 2, the dead boat rolled over and the men clung to the upturned bottom. Only the two most severely injured men were kept on it; Kennedy ordered the others into the water for fear they would be seen. The hull was spotted by Coastwatcher Reginald Evans on August 2, who radioed that he could see something floating but no sign of survivors. This was in response to messages from Rendova and other Coastwatchers to look out for 12 or 13 men.

What should they do? wondered the survivors. Wait until the hull sank completely and then swim? Or drift with the currents as long as possible, completely exposed?

Robert Donovan reports that there was a discussion. Kennedy asked whether the men, if found by the Japanese, preferred to surrender or fight. No one was keen on surrendering, knowing the Japanese reputation for cruelty to POWs. They elected to wait and see what happened before they made their choice.

The issue of whether to swim or drift, however, had to be resolved quickly. Although the swimming distances were daunting, refuges on various islands were visible, none more than five miles away. For a nonswimmer with burns, they looked unreachable. For a Harvard swim team competitor like Kennedy, they were possible.

Kolombangara was the closest at two and a half miles or so, but it had too many Japanese, and it was too far removed from the normal patrol routes of American ships. The other options included smaller atolls stretching five miles southeast of Gizo, which itself had a Japanese encampment and a seaplane base. The more inviting chain of tiny islands southeast of Gizo rose about three and a half miles from them.

The westernmost was Kasolo (Plum Pudding to English speakers). Others closer to Gizo were Bambanga (Long) and Naru (Cross). Some were too small to hide shipwrecked sailors. Others might have Japanese outposts. The group settled on Plum Pudding, not the closest, but judged the least likely to be occupied by the Japanese because it was so small. By this point the morning was dragging into afternoon. Floating on the sinking hull through another long night seemed less desirable to the survivors than making a break for it. Sharks were always a worry, though none had yet appeared.

The survivors inventoried their weapons. They had half a dozen pistols, three knives, a submachine gun and the 37-mm gun, now hanging overboard into the ocean. They had a lantern and a flashlight. No food, no water, no medicine, and no life raft, either—it had been taken off the ship to make room for the heavy gun. Plus, Pappy McMahon was in utter misery.

The crew did have a wood plank that had been used to mount the 37-mm. Gauging the distance to the atolls and the men's exhaustion, Kennedy and the group decided that nine of them, shepherded by Thom, would hang on the plank and kick and paddle for Plum Pudding.

Kennedy would independently pull Pappy doing the breaststroke, with Pappy's life jacket strap in his teeth, Pappy's back to his back. According to Donovan, Pappy was convinced that he had no chance for survival. But he let the younger skipper go about his business.

Sometime after one in the afternoon, the surviving *PT 109* crew began swimming for Plum Pudding. Neither passage was ideal, but the group wallowed along, Kennedy gulping down seawater, Lennie Thom acting like a tugboat for the plank barge of men. Not until dusk did the men reach Plum Pudding, and Kennedy actually beat the human barge to the shore. He was exhausted, so drained that Pappy lent him a hand to stand up. He promptly vomited up salt water on the deserted beach.

The crew of the *Amagiri* meets in a 1961 reunion. There was later some question as to whether contradictory orders were given by Captain Hanami (far left) and his superior, Captain Yamashiro, just before the collision.

Donovan guesses that they swam three and a half miles in four hours, a remarkable accomplishment considering how burdened they were. Kennedy himself would learn in the coming days that the ordeal could have been much worse had wind and current opposed them.

They had no illusions that they were out of danger. Just as they pulled themselves up on the beach, a Japanese barge came rolling by. The Americans thought they'd been seen, and dove to the sand and crawled up into the tropical underbrush. But the barge plowed on. They regarded the close call as a warning that their situation was precarious—they would need to be very careful not to give themselves away to the wrong side.

So, how were they supposed to get in touch with the U.S. Navy without being seen by the Japanese, who were everywhere? In this phase of the Solomons campaign, American air cover was still spotty and Japanese aircraft still roamed widely. The Americans could not yet deliver a knockout blow to the Tokyo Express or to the Japanese installations at Vila and Munda and on New Georgia. American dominance of the region would not come until well into 1944, when dozens of new aircraft carriers and thousands of new aircraft would come to the front lines. This was the reason that Kennedy had been running only one engine the night before, so as to avoid throwing up a phosphorescent wake and thereby giving away his position. The survivors of *PT 109* were huddling on a tiny spit of an island five miles from the Japanese installation on Gizo. Their home base on Lumbari was almost 40 miles away.

Kennedy decided that his only option was to try to get the attention of an American patrol. They passed these islands as they came through Ferguson Passage into Blackett Strait, just as his own squadron had done the night before. The question was, how? The survivors had no flares, they had no radio. They didn't even have a life raft. They merely had some pistols. And they were desperate men.

Kennedy decided to swim out into the passage to flag down an American ship.

Others of the crew thought that Kennedy was at best taking a huge risk, at worst making a huge mistake. But the issue was settled in a discussion between the three officers present—Kennedy, Thom, and Ross. Despite the reservations of the others, Kennedy decided to swim out into the strait with a flotation belt, a battle lantern, and a pistol, and once there to fire the pistol to attract the attention of American ships steaming past. At dusk on August 2, still exhausted from the swim to Plum Pudding the night before, he forged out into the Ferguson Passage.

He alternated swimming and walking between two and three miles along the reef that extends southeast from Plum Pudding. At times he was able to stand up, other times he was forced to swim. With night fully upon him, he plunged out into the open water.

He spent long, lonely hours out in the passage, swimming in an ocean that held all manner of perils, fatigued and growing chilled despite the warm waters, no American ships in sight. He would later learn that the PTs had been deployed to the north that night; they were nowhere near his swimming range.

But he didn't know that. He fired his pistol, the sound of which was swallowed up in the vastness. He continued treading water. Sometime in the wee hours of morning, he finally surrendered and began swimming back to the reef that would lead him to Plum Pudding and his men.

But the ease with which he had swum out was not to be repeated. The currents in Ferguson Passage reverse themselves with the change of tides, and Kennedy was caught on the wrong side of the current shift. When dawn came, shaking with exhaustion, he saw that he had hardly moved position. The current was pulling him farther away from the reef, not closer. Fearing he was close to drowning, Kennedy swam to minuscule Leorava Island, collapsed on the beach, and soon was asleep. His swim that evening is estimated at between five or six miles, although he covered less in distance due to the opposing currents.

Kennedy's failure to return to his men that dawn contributed to their gloom. Their skipper was lost, and they were far from being found. This latter belief was not inaccurate. Coastwatcher Evans and his network of native scouts had failed to spot any shipwrecked sailors.

Luckily, two teenage native scouts on the Allied side, Biuku Gasa and Eroni Kumana, were watching Gizo from a post on nearby Sepo Island. Although they didn't see Kennedy or his men, they observed as the Japanese landed several hundred troops even as Kennedy was floundering in the ocean on the night of August 2-3.

The two decided to travel by dugout canoe to Coastwatcher Reginald Evans's post on Kolombangara to report this news. They followed the island reef line before setting out across the strait. By coincidence, they found some floating debris—a shaving kit and a letter from Raymond Albert that had never been mailed—from *PT 109,* although they couldn't identify it as such.

They recovered the material and carried it to a colleague on an intermediate island. This scout, Benjamin Kevu, who could read English, took note of the letter,

Leaving the shattered hull of *PT 109*, Kennedy and the other survivors swam first to Plum Pudding (now Kennedy) Island and three days later to Olasana Island, where scouts Biuku Gasa and Eroni Kumana came ashore, as seen in this reenactment almost 20 years later.

Previous pages: This view, taken from Gizo Island, shows the smaller islands and Blackett Strait to the massive silhouette of volcanic Kolombangara. On Kolombangara an Australian coastwatcher in a secret post had spotted something in the strait, perhaps the hulk of an American PT boat.

although he did not yet connect it with the missing *PT 109*. Evans meanwhile received a report that an air search for survivors mounted the previous day had turned up nothing.

Because the air search hadn't been mounted until sometime in the afternoon of August 2, they missed the *PT 109* crew. By the time the planes flew over the drifting hull, the surviving crew had already disappeared into the bushes on Plum Pudding. Not sighting any survivors, Lieutenant Commander Warfield decided against sending a PT search party, for fear of putting another boat and crew at risk.

Although Kennedy didn't know the details of the rescue efforts made or not made on his behalf, he knew he couldn't let his men down. After sleeping for a time on Leorava, and despite being exhausted, he swam nearly two miles to Plum Pudding, where he arrived in late morning. As Kennedy emerged from the waves, one of the men mistook him for the missing Andrew Kirksey, in Donovan's account. Kennedy once again fell retching to the ground. He was dragged into the bushes to recover. While he was barely able to move the rest of the afternoon, he did manage at some point to wake up and command Barney Ross to attempt the same kind of swim that night.

Preparing for an epic swim, Barney Ross was not as confident or driven as young John Kennedy. He was worried about sharks, he wasn't as good a swimmer, and he didn't like the idea of being adrift in the black ocean all alone in the middle of the night. The only spark that kept him going was a desire to make the same sacrifice for his fellow crew members that Kennedy had made.

Ross left around four in the afternoon, so as to get farther out into Ferguson Passage before sunset. His swim into the passage was much shorter than Kennedy's had been. When he decided to give up the effort, he had no trouble in reaching Leorava Island, most likely because he was closer than Kennedy had been and the tide was not running against him. After sleeping on perhaps the same beach that Kennedy had used the night before, Ross swam back from Leorava Island on the morning of August 4. The men were glad to see him return, and he was in good spirits after having survived his perilous adventure.

Also on August 4, the two native scouts, Biuku and Eroni, left Kolombangara to return to their home base on Sepo Island, ordered once again to look for signs of survivors. For a second time, the *PT 109* crew and the islander scouts would be in fairly close proximity to one another but would not make contact. The scouts reached the island of Wana Wana, where they spent that night with other scouts and then continued on to Sepo on August 5.

August 4 was not a good day for the survivors. Morale among the enlisted men was sinking ever lower. Attempting to keep up the spirits and loyalty of his men, Kennedy had to fight down his anger at the Navy for not sending out a search party.

In addition, the lack of water and food had become a serious problem. They had gone three days since their last meal. Plum Pudding, they found, didn't have enough coconuts to sustain them. The next island down the chain to the southeast, Olasana, was thickly forested. Its size and tree cover triggered concern that it might harbor Japanese, but the men had seen no Japanese activity during their time on Plum Pudding and the decision was made to move to the bigger island. It was a gamble.

The surviving crew of *PT 109* swam as they had when evacuating the sinking hull, Kennedy pulling Pappy by his kapok strap, back to back, while doing the breaststroke. The others kicked along on their salvaged wood plank. The distance was 7,000 feet as the crow flies, but the currents swirling through the smaller islands probably increased the distance. They swam for hours before making landfall on their new home.

Though the island had a more generous supply of coconuts, its relatively larger size and dense undergrowth made the Americans hesitant to go exploring lest they encounter a Japanese party. The men remained hidden in an area about 50 yards by 50 yards. According to Robert Donovan, one of the coconuts they harvested made Kennedy and some of the others sick. Barney Ross ate a live snail, more a fraternity prank than serious effort at nourishment.

They were now stationed near the southeast end of the chain of small islands stretching from Gizo to Ferguson Passage, too fatigued from lack of food and the day's long swim to think of venturing out into the passage to hail a passing PT boat. For the first time in two nights, the American PT boats were again deploying into the strait. Had Kennedy or Ross been out in the passage around nine o'clock, or again toward dawn, they might have been able to hail their passing comrades.

Coastwatcher Evans, observing the Strait from his vantage point on the flanks of Kolombangara, radioed reports regarding the hull of a small vessel seen floating in the area. By the evening of August 4, messages had been exchanged ordering Evans to attempt to destroy the wreck before the Japanese could get to it. As the wreck shifted position, it became harder and harder for Evans to see how he could accomplish the task.

On the morning of August 5, Kennedy and Barney Ross decided to swim to Naru, the last island in the chain before the open strait. They were hesitant, because although Naru provided a great vantage point onto Ferguson Passage, it was likely the Japanese had set up an observation post there.

Kennedy left the crew under the command of Lennie Thom. He and Ross swam the easy 4,000-foot swim to Naru. They found it seemingly deserted. They crossed over to the shore facing the passage and made discoveries that both instilled fear and offered them a shred of hope.

First, out on the reef toward the Passage they spotted the wreck of a small Japanese vessel that had been destroyed by bombing, perhaps the same bombing that Kennedy and the others had watched from a distance while on Plum Pudding.

On the beach they found a crate of Japanese candy. It had either washed ashore from the bombed vessel or been dropped overboard by a Tokyo Express run and drifted to the little islet. Kennedy and Ross ate some of the candy and pondered how to get it back to their hungry men. Farther down the beach, the two men found a hidden dugout canoe and a supply of drinking water. These were native scout caches, although Kennedy and Ross didn't know that. They drank some of the water and wondered what to do with the canoe.

The different players in this drama were beginning to converge on one another. Finally, Biuku and Eroni left Wana Wana on August 5, headed for Sepo. As they crossed from Wana Wana to Naru, they were intrigued by the same Japanese wreck Kennedy and Ross had just observed and they pulled in closer to inspect it. Ross and Kennedy, having satisfied their thirst, stepped back onto the beach. The two pairs of men saw each other and both were frightened. Kennedy and Ross dived into the bushes, and Biuku and Eroni paddled away as fast as they could. Biuku describes what happened next this way: "We went over to have a look. We thought we could find some clothing or food on the wreck. We came around on the calm side, anchored the canoe. We swam up and went inside the ship—lots of things were in there, like rice, clothes, stacks of things. It was only half of the ship—the bow half. I took a coat and Eroni took a bayonet. Then Eroni saw a white man walking along Naru. We didn't know whether he was Japanese, but we said, 'Let's go.' We took what we could, swam down, and left."

Kennedy and Ross feared the natives would give them away to the Japanese. Eroni and Biuku feared the two men they'd seen across a mile of open water were Japanese, and that their lives were in danger. At least the two islanders had the luxury of an escape. All the Americans could do was hide in the undergrowth.

The islanders normally would have continued on to their home on Sepo. But following this first brief encounter between the islanders and Kennedy, Biuku persuaded his friend to change course for Olasana, where they could pick up some fresh coconuts. He was thirsty from the hot paddle away from Naru and needed a drink.

Coastwatcher Lt. Arthur R. Evans (above, third from right) received word in person from the scouts about the Americans' situation. He sent other scouts with a message (below) to the castaways: "... Have just learnt of your presence ... and also that two natives have taken news to Rendova.... " Rescue was under way.

Lieutenant Kennedy carved this message on a coconut shell and sent it with
scouts bound by canoe for the PT base at Rendova: "NAURO ISL / NATIVE KNOWS
POSIT / HE CAN PILOT   11 ALIVE NEED / SMALL BOAT / KENNEDY." When John
F. Kennedy was President, the coconut sat on his desk in the Oval Office.

Eroni agreed. But for that detour for coconuts, Kennedy and the others might have slowly starved on Olasana, been shot during a paddle attempt for Rendova, or otherwise fallen into Japanese hands.

The two young natives, neither yet 20 years old, paddled toward the shore. On Olasana, Thom and the enlisted men crouched, watching the natives approach. A whispered discussion ensued—were they scouts for the Japanese on Gizo? Was this the end?

Lennie Thom took a major gamble. He walked onto the beach to try to communicate with the native boys.

The two scouts were understandably frightened, suspecting the worst about the strangers—that they were the enemy—and paddled back out into the water. Thom beckoned to them to return, calling, "Come, come," but they were unconvinced. He shouted things like "Navy" and "Americans," but the two islanders didn't speak English well enough to understand. Thom called out the name of a native scout he knew, but they didn't recognize it. Finally, he said "white star" and pointed to the sky. The islanders knew that the white star referred to American insignia on the planes. They slowly returned to shore to greet these curious-looking Americans, their allies. They had been told by their Coastwatcher leaders to rescue and help any downed "white star" pilots, and to deliver any surviving men to the Coastwatchers.

Biuku tells it this way: "We went to Olasana. I was thirsty, so I told Eroni to leave me at the island so I could get a coconut. As I was going ashore, I saw a white man crawling out from the bushes hear the shore. I said, 'Eroni, a Japanese here,' so we pushed the canoe out to get away. We were headed out when he stood up, waved, and said, 'Come.' I said, 'No, you're Japanese.' I think he understood me, because he said, 'I'm an American, not Japanese.' He showed his arm and said, 'Look at my skin. It's white. Japanese skin red.'"

Convinced, they pulled their canoe onshore and the enlisted men helped hide it in the undergrowth. The two groups fumbled at communicating across the cultural and linguistic chasm that separated them, but finally several ideas got through. The islanders logically but mistakenly claimed that Japanese were on Naru, because they had just seen the wreck on the reef and then Kennedy and Ross out on the beach. Thom and the crew mistakenly concluded that Kennedy and Ross were in great danger.

Two separate efforts were made to contact the Navy and rescue the men on Naru from their predicament. Thom, Starkey, and Biuku attempted to take a canoe to Rendova that afternoon, a distance of 38 miles, but made it only into the Ferguson Passage before heavy seas drove them back. They returned just as Kennedy arrived back at Olasana with the Japanese candy. The men were overjoyed to see him. He met the two islanders who had spotted him on the beach in Naru, but neither side realized they had already run into each other.

Kennedy then returned to Naru and commanded Ross to accompany him on a canoe paddle out into the strait to hail the PT boats. Ross thought the water was too rough, but they went out, only to capsize beyond the reef. They swam back toward shore, and in a rush of surf were tossed back over the reef into the coral bed onshore, miraculously avoiding serious injury. PT boats passed through the strait that night, but no *PT 109* adventurers were out there to catch their attention.

As this drama was being acted out on Naru and Olasana, Coastwatcher Evans had concluded that he would need to move his camp to the other side of Blackett Strait in order to better observe the activities in the strait. He felt his vantage point on the side of Kolombangara was too removed from the major operations. He elected to move to the island of Gomu, about seven miles south of Naru, on the afternoon of August 5.

Following their attempts at swimming and sailing for rescue, Kennedy and Thom independently decided to try to send written messages via the native scouts. Kennedy had Biuku open a coconut for him, and carved a brief message to the PT commander in Rendova:

NAURO ISL
NATIVE KNOWS POSIT
HE CAN PILOT   11 ALIVE NEED
SMALL BOAT
KENNEDY

Thom wrote a more detailed message in pencil on a blank invoice from a Gizo trading house he happened to have.

On August 6, both messages went off with Biuku and Eroni, who set out for the long paddle to Rendova. They stopped first at Raramana on the island of Wana Wana, where they told Benjamin Kevu, the English-speaking native scout, about the survivors. Kevu dispatched another scout to Gomu to wait for Evans to tell him about the survivors on Olasana. On Wana Wana, Biuku and Eroni picked up another scout to help with the long paddle south.

Evans arrived at Gomu on the night of August 6. He sent another canoe with scouts back to Olasana acknowledging that the American survivors' message had been received and confirming that a canoe was being sent back to pick up the senior officer and bring him out.

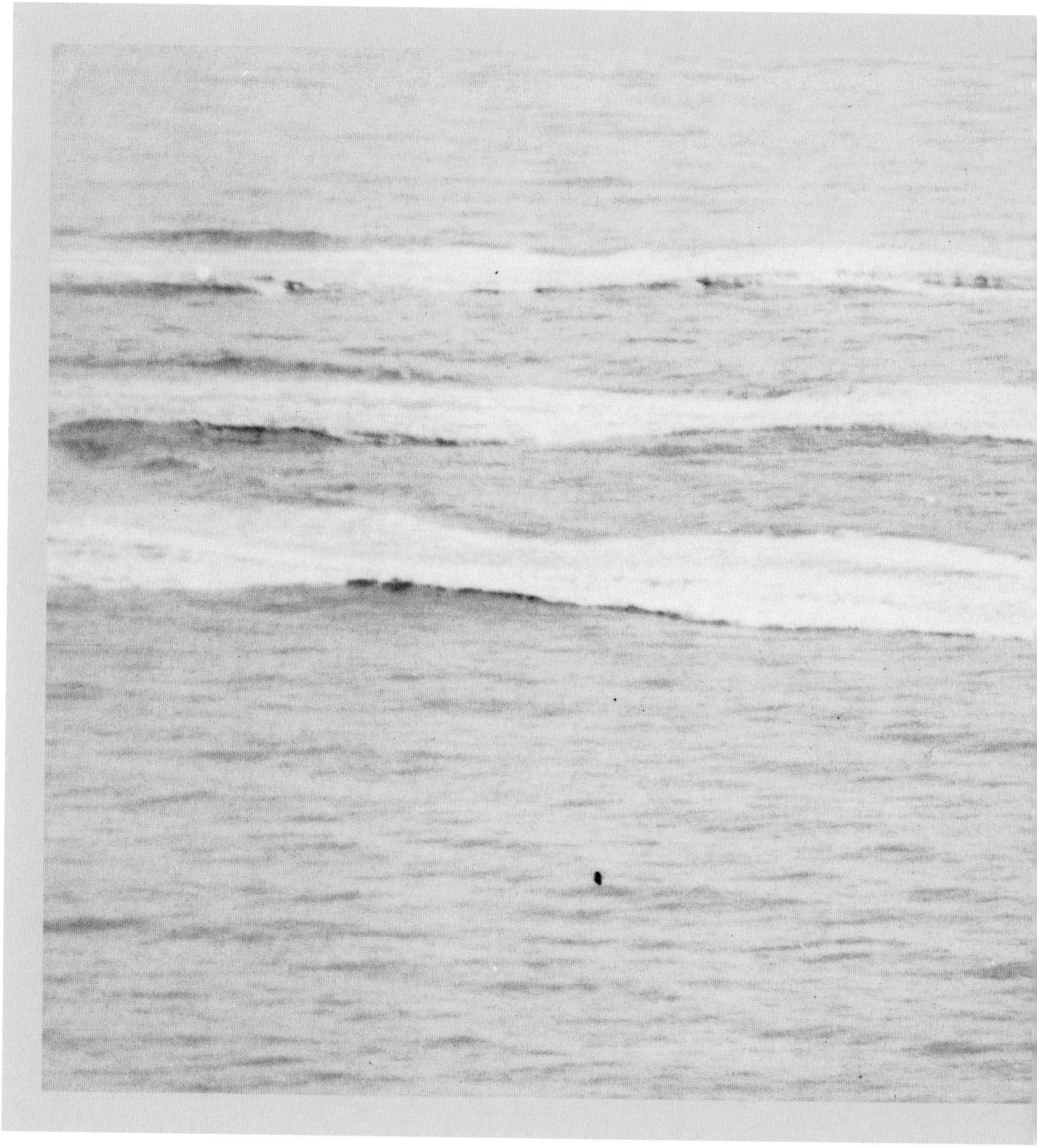

some advisers. And 40 of those advisers were telling us that they wanted to attack the Vietcong, and that if they were allowed to do so, they were sure they could wipe out the Communists. Jack's response was to warn them that if they did, not only would those 40 American advisers be pulled out, but 40 more. He was convinced that we couldn't win Vietnam militarily; he wanted to do it diplomatically, and I think he came to that conclusion because of the war in the Pacific."

Blunt-spoken Dick Keresey thinks Kennedy had very strong and controversial opinions about his experience, but he didn't share them with many people. An exception is an exchange that Keresey says took place between Kennedy and his fellow PT commander Potter on the beach at Lumbari the day after the rescue.

"A friend of mine, Dave Payne, saw the two of them on the beach. Jack was feeling good enough to be up and around. And Kennedy was furious with Potter, and was letting him know it. You could just see Potter growing cold as Kennedy attacked him. I've never heard a conversation like that in the Navy."

Presumably Kennedy was venting his anger about the lack of a rescue attempt, the fault for which he believed was shared by Brantingham, Lowery, and Warfield. The issue drove a wedge between Kennedy and those he had expected to save him, one that lasted the rest of his life.

The first letter Kennedy wrote to his family after the collision shows no significant change in thought, other than a renewed desire to be home. The letter—which passed through Navy censors—exhibits a new respect for American courage and American fighting men. And a new fatalism about his lost crewmen comes through, as well as his regret about the lives that war consumes.

Letter received from Jack, September 12, 1943:

Dear Mother and Dad:

Something has happened to Squadron Air Mail—none has come in for the last two weeks. Some chowder-head sent it to the wrong island. As a matter of fact, the papers you have been sending out have kept me up to date. For an old paper, the New York Daily News is by far the most interesting.

I saw where Zeke's cousin, Chuck Spaulding, who roomed with George Mead, has published his book "Love at First Flight." It's dedicated to George and is the story of his (Chuck's) pre-flight training. It's supposed to be excellent. Would you send it out?

I see where the show is going to be produced. I'm sorry it isn't going to be a musical. I'm just about ready to come back and laugh for a week with a lot of chorus girls. I also saw where Bunny Waters had received a diamond bracelet from an unknown admirer—who one week later in the same column, turned out to be Pal Jolsie. I guess her heart still belongs to Mammie. Bunny sent me her picture with a long inscription and she is now my No. 1 pin-up, now that Miss Angela Green's picture is buried deep in the heart of Ferguson Passage.

In regard to things here—they have been doing some alterations on my boat and have been living on a repair ship. Never before realized how badly we have been doing on our end and although I always had my suspicions. First time I've seen an egg since I left the States.

As I told you, Lennie Thom, who used to ride with me, has now got a boat of his own and the fellow who was going to ride with me has just come down with ulcers. (He's going to the States and will call you and give you all the news. Al Hamm). We certainly would have made a red-hot combination. Got most of my old crew except for a couple who are being sent home, and am extremely glad of that. On the bright side of an otherwise completely black time was the way that everyone stood up to it. Previous to that I had become somewhat cynical about the American as a fighting man. I had seen too much bellyaching and laying off. But with the chips down—that all faded away. I can now believe—which I never would have before—the stories of Bataan and Wake. For an American it's got to be awfully easy or awfully tough. When it's in the middle, then there's trouble. It was a terrible thing, though, losing those two men. One had ridden with me for as long as I had been out here. He had been somewhat shocked by a bomb that had landed near the boat about two weeks before. He never really got over it; he always seemed to have the feeling that something was going to happen to him. He never said anything about being put ashore—he didn't want to go. He never said anything about being put ashore—he didn't want to go— but the next time we came down the line I was going to let him work on the base force. When a fellow gets the feeling that he's in for it, the only thing to do is to let him get off the boat because strangely enough, they always seem to be the ones that do get it. I don't know whether its just coincidence or what. He had a wife and three kids. The other fellow had just come aboard. He was only a kid himself.

It certainly brought home how real the war is—and when I read the papers from home and how superficial is most of the talking and thinking about it. When I read that we will fight the Japs for years if necessary and will sacrifice hundreds of thousands if we must—I always like to check from where he is talking—it's seldom out here. People get so used to talking about billions of dollars and millions

"For heroism … " began the citation. Captain Frederick L. Conklin
pinned the Navy and Marine Corps Medal on Lieutenant Kennedy in
1945 for " … courage, endurance and excellent leadership … in
keeping with the highest traditions of the United States Naval Service."

In 1944, a reunion brought together *PT 109* crew, wives, and family at Hyannis
Port (opposite, above). Competition included the inevitable touch football game
(opposite, below). A replica of *PT 109* (above) passed the reviewing stand
at the 1961 Inaugural of President John F. Kennedy.

PT 59, right, ran on its most dangerous mission under the command of Kennedy:
to evacuate a platoon of surrounded Marines who had been sent to Choiseul Island
in a diversionary feint. On half fuel, *59* made the rendezvous under Japanese fire,
took the Marines aboard, one of them Dick Keresey, and returned to base.

On the morning of August 7, Biuku and Eroni decided to deliver their message to the Navy base on Roviana off Munda, since it was much closer than Rendova. Now everyone in the area was informed of the discovery of the survivors. Meanwhile Evans sent a canoe with the cream of his island scout team, including Benjamin Kevu, to keep the survivors comfortable and to transport Kennedy back to Gomu.

Olasana was large enough and the Americans were so well hidden that this larger native party had a hard time at first, but the party finally found the survivors and served them a sumptuous meal of potatoes, rice, fish, cigarettes, and C rations. They built a hut for Pappy McMahon, and offered up fresh coconuts.

Shortly after dinner, Kevu asked Kennedy to accompany him to Gomu. There was no more room for the other Americans in the rescue canoe; the crew would have to wait until Kennedy and the rescue party returned. Kennedy was amused at the formality of Evans's written message introducing himself and his men, and for the first time in a week, spirits rose. Even Pappy McMahon began to sense that their ordeal might have a happy ending.

Kennedy and his escorts were halfway across the strait when a Japanese plane buzzed them. After much debate over whether to respond, Kevu stood up and waved a friendly greeting. The gesture was apparently enough, for the plane flew on without harassing them.

In Gomu, still dressed in his underwear and matted with the rime of a week's swimming and coral injuries, Kennedy met Coastwatcher Reginald Evans. Through Evans, Kennedy persuaded the Navy to allow him to meet and join the PT rescue party at an island called Patparan. He would then lead them to the hidden men.

The rendezvous with the American PT boats occurred about ten that night, Kennedy firing the agreed four shots from a captured Japanese rifle. He joined the Americans, who included skipper Al Liebenow of *PT 157,* Alvin Cluster, Edward Brantingham, Biuku, Eroni, and several American news correspondents.

Two hours later, with Kennedy guiding them, this larger group used a rubber raft to penetrate the reef around Olasana. After midnight on August 8, the survivors of *PT 109* were loaded from the northeast corner of Olasana onto *PT 157* and taken back to Lumbari, where they were greeted as returning heroes.

THE HEROISM OF KENNEDY AND OTHERS OF HIS CREW in saving the lives of the *PT 109* crew is not in question. As to whether Kennedy was negligent in getting rammed or in how he rescued his men, most experts give him the benefit of the doubt. War is fought in real time, not in hindsight, and snap decisions can be second-guessed as long as there are those willing to do it.

"A collision like that could have happened to any one of us out there," says Dick Keresey. He attributes most of the second-guessing of Kennedy's skills and actions to Potter, Lowery, and Brantingham, motivated by the bad feeling that lingered between them and Kennedy over their failure to mount an effective search and rescue mission that first night.

Historian Richard Frank says that Kennedy as a skipper was "about average under the circumstances, and I think that's how he was judged by his peers. And his conduct afterward, I think the finest testimony to that is the way his crew regarded him. They revered him for his conduct, and they were the judges best placed to assess what he was really like."

The fact is, none of the PT commanders distinguished themselves that night in the Battle of Blackett Strait, according to Robert J. Bulkley. "This was perhaps the most confused and least effectively executed action the PTs had been in—the chief fault of the PTs was that they didn't pass the word. Each attacked independently, leaving the others to discover the enemy for themselves."

One observer who thinks Kennedy did his best in difficult circumstances is analyst and former Army lieutenant Dale Ridder. Some people, he says, point to the fact that *109* was the only PT ever sunk by ramming as proof that Kennedy screwed up. "True, no other PT was rammed and sunk, but it came very close to happening to two boats at the same time. As for whether Kennedy erred by having the crew leave the hull and swim ashore, there were quite a few friendly-fire incidents involving aircraft and PT boats at this time. In my view, an attempt by the crew to signal a "friendly" aircraft would have resulted in them being shot to shreds. As for a daylight rescue by Catalinas, [look] what happened when the Marines tried a daylight pickup of wounded in the Rice Anchorage area by Catalinas."

Ridder is also forceful in pointing out the difficult visibility conditions Kennedy and the other PT boats were operating under that night and most nights in Blackett Strait: "I used one of my planetarium programs, 'Starry Night Deluxe,' to determine the moon status on 2 August 1943 at 0200. At that point in time the Moon was on the opposite side of the Earth, so there was no moonlight whatsoever that night. Having been on night patrol exercises while in the Army, I do not think that the average person understands how limited the visibility is under those conditions.

"As to whether the crew were somehow negligent in getting themselves rammed, you should look over the information about the first experiences of the Guadalcanal

Kennedy relaxed with officer pals Jim Reed, Barney Ross, and Red Fay (below). His impish *PT 59* crew (above) kept after him to blast an enemy target— an abandoned latrine. Finally Kennedy agreed and got permission to fire target practice. Robert J. Donovan wrote: "The last remnants of Hirohito's empire on that beach went up in smoke and splinters."

boats, especially the case where one destroyer almost rammed two boats at once. As for trying to keep in touch with a boat 2,000 yards away, that was impossible. See the PT reports on the attack on the *McCawley*, where the transport, a 10,000-ton ship, was detected at 800 yards. Keeping in touch at night for PT boats before radar was essentially impossible unless they were on top of each other."

Dick Keresey disputes whether Barney Ross had night blindness at all. "I never heard that," he says. As for who was to blame for the crash, he puts the onus on the shoulders of Warfield and his colleagues at Rendova for ordering back to base the radar-equipped lead boats after they had fired their torpedoes.

Whatever the cause of the sinking, Kennedy and his men were not thinking about postmortems when they arrived back at Rendova. The war still needed to be won, and Kennedy wanted a chance to get even with the Japanese. He was also furious that they had not been rescued for a week.

In his 1966 bestseller *The Pleasure of His Company*, "Red" Fay recalls how after hearing of the collision and sinking of *109*, he was most personally moved over the seeming loss of Barney Ross. But he was particularly angry that no effort had been made to find Kennedy and crew. He confronted another PT commander, Philip Potter of *PT 169*, who had been in the area with Kennedy on the fateful night, and asked if Potter had really conducted a search after the collision. When Potter said yes, Fay accused him of lying.

"Red" Fay only became friendly with JFK after the rescue, when Kennedy was recovering from his ordeal at Tulaghi in August 1943. Red was part of a small flotilla composed of LCI, LCT, and PT boats.

"We were on our way down to get the PT worked on, and 12 Japanese torpedo planes jumped us," Fay narrates. "The crazy thing was, the Japanese torpedoes were hitting the surface and skipping along the surface like stones, and when one finally hit us it didn't detonate, but plowed into the hull, leaving a hole but causing no explosion. I went down below with an engine mate to check on the torpedo. The other guy said he couldn't tell if it was armed or not.

"We later heard from Tokyo Rose that the Japanese were claiming they had hit a cruiser and two destroyers. That gave us a laugh.

"At this point," says Fay, "the skipper decided to abandon the LCI, when it really didn't seem necessary. He threatened to use his .45 to enforce the order. Another guy with no shirt on demanded that we not abandon, nor that the skipper touch his gun. The skipper said he would court-martial the guy for insubordination. As it turned out, we got back safely to Tulaghi and the skipper was court-martialed, while the other guy got a Silver Star for saving the ship. When I was Navy undersecretary some years later, I ran into him in a reception line in Florida. I recognized him and we became friends. We're still in touch."

At Tulaghi in late summer 1943, life began to settle back into a more pleasant routine for Kennedy.

"Jack was a terrible poker player, so he read while we played," says Fay. "If he had played he would have lost all his money." But what impressed Fay more was that Kennedy supplied the other officers with a steady stream of magazines, like *Time* and *Saturday Evening Post*. And he made the other officers read and think. "Jack was always warning us that if we didn't get a better understanding of why we were in this war, we'd be in another one sooner than we wanted." Kennedy also led the men in serious bull sessions, according to Fay, about politics, foreign policy, military strategy, and even education.

Another incident further increased Fay's estimation of Kennedy. A young group of enlisted men arrived on Tulaghi, fresh from the U.S., ready for their first combat duty. Someone needed to orient them to the rigors of life on PT duty, and Kennedy volunteered even though he was still weak and gaunt from his brush with death a few days before. According to Fay, the new men gathered close around Kennedy because he spoke informally to them rather than like a military briefer at a podium, and he really held their attention. After seeing that, says Fay, "I thought this guy had a chance to be president. I put the odds at one in 10,000 and for the rest of us at one in a million. Not so much because of his family background as because of his seriousness and his communications skills."

After the afternoon and evening bull sessions, the young officers would go to the officers club for a drink and then dinner. Fay recalls that the dinners were pretty good, but when pressed to name the entrées, all he can remember is Spam. "It tasted pretty good back then," he says, "and we took a lot of it on board because it would keep in the heat. Now I can't eat the stuff."

Asked whether he thought the experience of *109*'s sinking affected Kennedy's personality, Red Fay says no. "Jack had a strong personality before and after. I don't think he was changed by the event." But he believes some of JFK's views were influenced by his experiences in the Solomons, particularly with respect to Vietnam. "I was with him while we were sailing on *Honey Fitz*," he recalled, "and a cable made its way to us from the U.S. forces in Vietnam. All we had in Vietnam at that point were

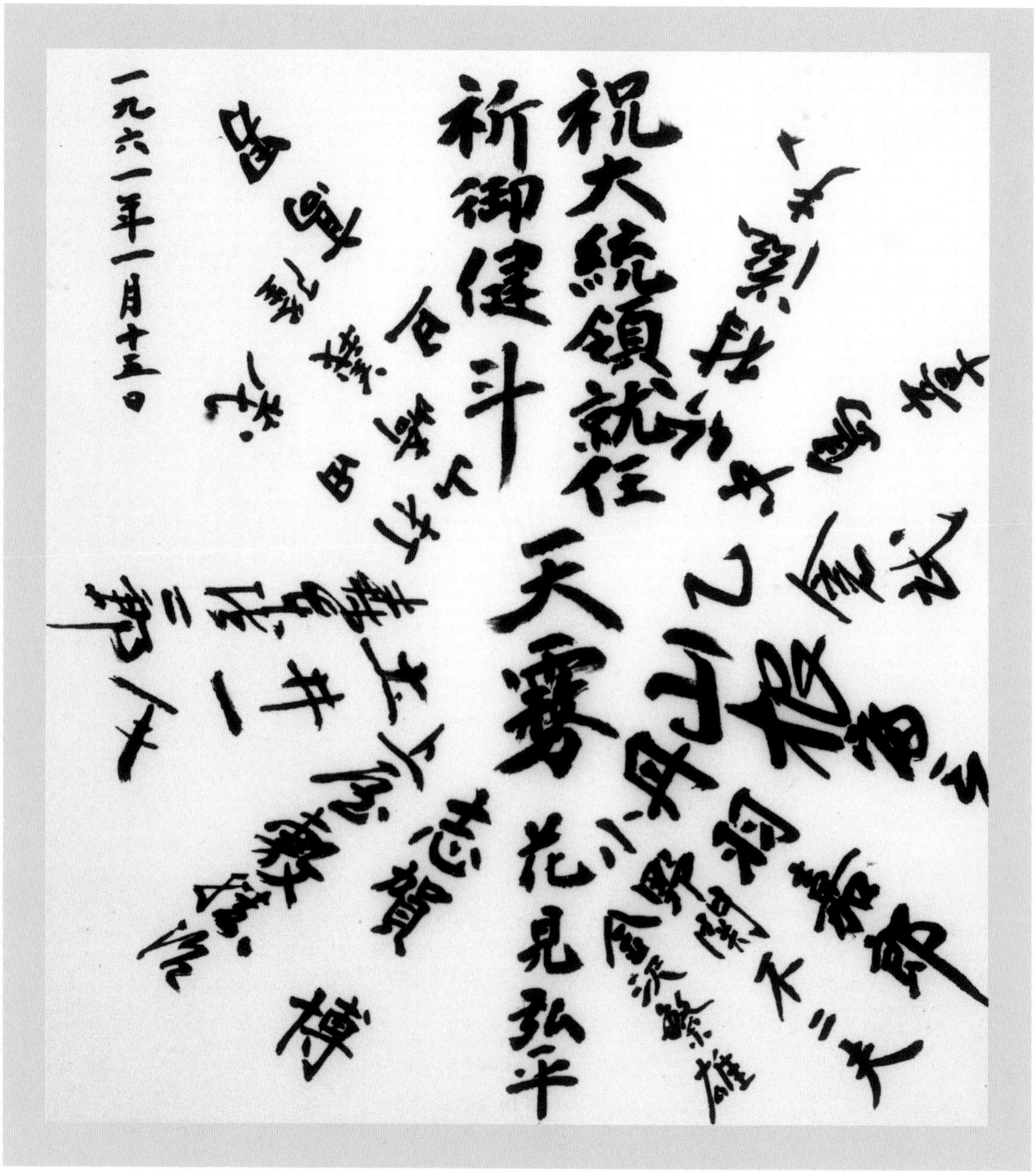

祝大統領就任
祈御健斗

天霧

一九六一年一月廿日

花見弘平

"Sincere congratulations" and "best wishes for success" were expressed alongside the signatures of the surviving captain and crew of destroyer *Amagiri* of the Imperial Japanese Navy in a greeting sent on the occasion of John F. Kennedy's Inauguration.

of soldiers that thousands of dead sounds like drops in the bucket. But if those thousands want to live as much as the ten I saw—they should measure their words with great, great care. Perhaps all of that won't be necessary—and it can all be done by bombing.

Has Joe left yet—I hope he's still around when I get back. Saw Jake Pierrepont the other day who had received a letter from Marion Kingsland (of Palm Beach) who reported Joe in New York with two of "the most beautiful English girls she had ever seen." I hope, if Joe is planning to leave, he will leave a complete program with the names and numbers of the leading players.

We have a new Commodore here—Mike Moran—former captain of the Boise—and a big harp if there ever was one. He's fresh out from six months in the States and full of smoke and vinegar and statements like—"it's a privilege to behave and we would be ashamed to be back in the States"—and we'll stay here ten years if necessary. That all went over like a lead balloon. However, the doc told us yesterday that Iron Mike was complaining of headaches and diarrhea—so we look for a different tune to be thrummed on that harp of his before many months.

Love, Jack

PS. Got camera and Reading glasses. Thanks. Summer beginning and it's getting hot as the devil hence letter blurred. If you should see Mrs. Luce would you tell her that her lucky piece came through for me. I understand she has five of them herself. At their present rate of luck production, there is no telling where it will all end.

Once recuperated, Kennedy was eager to obtain another command. On September 1 he took *PT 59*. He was eager for the assignment because *59* had been retrofitted as a gunboat—and as Dick Keresey and others emphasize, the PTs would prove to be most effective as gunboats. On *PT 59*, John Kennedy would once again risk his life to save others, and a major role would be played by Dick Keresey. This time the drama took place in November 1943, on the island of Choiseul.

"Although we didn't know it at the time," says Keresey, "I was chosen to take part in a diversionary operation on Choiseul. I was sent with the 2nd Marine Parachute Battalion to establish a beachhead and a possible PT anchorage on Choiseul. The whole point of this, as it turned out, wasn't to take Choiseul; it was to divert Japanese attention from the coming U.S. assault on Bougainville, which was another Japanese stronghold."

Keresey spent a week with the Marines while they attacked the Japanese strength on the western end of the island. In one sector suddenly several Marine platoons were getting as good as they gave. The Japanese fire intensified, putting the Marines in a dangerous position. The Marine plan was to use LCVP boats to pick up the men and pull them away to safety, but Keresey suggested they use PT boats instead. "With the Japanese coming on," says Keresey, "I suggested that we bring in some PT boats to help the Marines and get them out. The word went up to Vella Lavella, where Kennedy and I were based."

After helping repair one of the LCVPs, Keresey saw them head off to rescue the endangered platoons. After sundown, he boarded one of the two PT boats that had finally come from Rendova and was surprised to meet Jack Kennedy once again. Kennedy was on *PT 59*. Apparently the order had gone out before he could refuel and his tanks weren't full. But he knew that to get in line and fill up with fuel would delay him by another half hour, so he went without filling up.

Because of this instinctive decision, criticized by some, he got to the action when he was needed.

Ten miles south of the Marines, Keresey was picked up by Kennedy, and he guided Kennedy's boat into the action. The Marines were still in jeopardy because the repaired LCVP had stalled out and was drifting toward shore.

"I guided Kennedy to the beachhead," says Keresey. "He used the sound of voices to find the endangered Marines. What he found was a group of men who had barely survived—many others had not—a harrowing encounter with a more powerful Japanese force. We off-loaded men and wounded onto Kennedy's boat from the stalled LCVP. Luckily when we performed this we weren't under fire. We got them back to Vella Lavella. This was the last time I saw Kennedy in the Solomons."

Warner Brothers tells the story more dramatically. The fleeing Marines are under fire and it is Kennedy's fuel-starved boat that drifts toward shore before it is towed away.

At the end of 1943, Jack Kennedy was sent home for a long-needed rest. Reports of his heroism in rescuing his men had already been heralded back in the U.S. in wire service reports and magazine articles. His life shifted even more dramatically in 1944, when his older brother's, Joe Jr.'s, plane was blown up and he was killed while on an "Aphrodite" mission. John F. Kennedy, who some thought would have preferred a life as a teacher or a writer, was suddenly the Kennedy family scion. His rise in politics was meteoric. By 1946 he was a winning candidate for Congress from Massachusetts. He became a senator and married socialite Jacqueline Bouvier in the 1950s. In 1960 he was

In the White House Oval Office, President Kennedy points out his position on the bridge of *PT 109* when she was rammed by the Japanese destroyer. He was thrown backward in the collision and aggravated old back injuries that became a chronic reminder of his wartime service.

elected President, and in November 1963 he was assassinated. His brothers Robert and Edward, who had visited with some of the *PT 109* veterans at Hyannis after the war, also became senators, and candidates for President. Robert was also assassinated in 1968.

The subsequent lives of the other *PT 109* survivors were not so newsworthy, but Jack Kennedy stayed in touch with many of them, particularly at the end of the war and the years immediately after. On his way back east in 1944, he visited with Pappy McMahon's wife in California. After the war, McMahon recovered from his burns and became a mail carrier in California. He and others, such as Barney Ross, worked on various Kennedy political campaigns, although the contacts grew less frequent during the 1950s as Kennedy climbed the political ladder. Ross, John Maguire, Charles Harris, Ray Starkey, Bill Johnston, Gerard Zinser, and Edgar Mauer even rode on a *PT 109* float in the 1961 Inaugural Parade.

PT veteran Red Fay became Undersecretary of the Navy, William C. Battle became Lieutenant Governor of Virginia, and Byron White went on to become a Supreme Court justice.

Did the collision of two ships in the Blackett Strait in August 1943 cause all those later events to come to pass?

We will never know the answer, but what we can do is look for surviving physical evidence of the event that happened long ago and perhaps separate fact from fiction. In this case, we must journey back to the dark-jungled Solomon Islands, to see what, if anything, remains of *PT 109*.

THE

ISLANDS

A woman and child rock gently while another child looks on in the village of Vila, once a Japanese base on Kolombangara. The Solomons were named even before discovery by 16th-century explorers in search of the legendary land that supplied gems, silver, and gold to biblical King Solomon.

TO THE PRESENT

THE SOLOMON ISLANDS ARE AN UNLIKELY SPOT for the turning point of a global conflict. On the ground in Gizo, a downpour drenches in what is supposed to be the dry season. The storm that hung out over the strait when we landed has caught up to us and is emptying its full force on this little island. Maybe we'll have to do our work in rain, not in blistering sunshine. I'm reminded that there is no patch of ocean and earth quite like this anywhere else in the world. The Slot on a calm day is nothing more than a giant lake in the ocean, which is part of the reason men settled here in the first place, and why Japan and the U.S. ended up dueling it out here in 1942-44. The islands served as giant aircraft carriers, and the smooth seas were like a conveyor belt, first for Japanese troops and supplies heading southeast to try to create a South Pacific "Maginot Line," and later for the U.S. men and their supplies headed northeast toward the Bismarck Barrier and beyond to the Philippines and Japan.

The islands result from the collision between two giant tectonic plates, the Oceanic and the Indo-Australasian. These two plates grind together, making for especially frequent seismic activity, particularly in Papua New Guinea and Vanuatu. The Solomons often record more than a thousand seismic movements per year, including dozens above 5 on the Richter scale. A big quake in Guadalcanal in 1977 measured over 7 and caused landslides and extensive damage. But the big quakes and tidal waves more common to Indonesia and Japan are rare here.

The name Solomon Islands is derived from biblical lore, and the country's boundaries are arbitrary lines drawn on a map, since geologically these islands are part of the long chain of volcanic mountaintops, peripheral reefs, and atolls that run east off the northern coast of Papua New Guinea, including the Bismarck Archipelago. And even the islands are not static entities; what are now scattered islands were once part of much larger landmasses that in turn were connected to a much larger New Guinea. The global warming and massive rise in sea level at the end of the last Ice Age in 10,000 B.C. reduced the island land areas to what we see today.

The source of the chain is volcanism, even though the uplifts have been adorned with later additions of coral reef, jungle, and beach, and many of the smaller outlying and intermediate islands are atolls and reefs, not seamounts. The true volcanic peaks of the Solomons include Savo off Guadalcanal, Kolombangara just opposite where the *PT 109* sank, and Simbo in the west. An undersea volcano known as Cook rumbles off Vonavona not far from where Kennedy fought, and another called Kavachi boils beneath the oceans off New Georgia.

Papua New Guinea is the second largest island on Earth, after Greenland, and it is fitting that its Solomon offspring are also large. The total landmass of the chain is 10,985 square miles. The largest of the 992 islands in the Solomons—Guadalcanal, Malaita, New Georgia, Santa Isabel, San Cristobal, and Choiseul—are comparatively large compared with those in other Pacific island groups in Micronesia and Polynesia farther north and west. Some of them exhibit characteristics more akin to continents than little islands, having high mountain peaks, dense jungle, plateaus, long flowing rivers with freshwater wetlands, and soils that include clay, igneous, and sedimentary components.

The human makeup of the island chain is just as much the product of collision and turbulence. Behind the current facade of pristine beaches and quiet villages are millennia of exploration, migration, tribal conflict, and colonization. Although the precise origins of the Solomon islanders are lost, it seems that beginning tens of thousands of years ago nameless groups made their way ever southeast from mainland Asia, into Indonesia and New Guinea, using the lower sea levels of the ice ages to follow land links that no longer exist, or to cross shallow gaps that are now deep gulfs.

By about 2000 B.C., the Solomons were most likely inhabited by these unknown Austronesian wanderers. The anthropological record becomes clearer with the arrival of the Lapita people. A distinct group named after an archaeological site to the northwest in New Caledonia, the Lapita came ashore in the Solomons around 1600 B.C. and they are recognized by their pottery and its derivatives, which can be found throughout

Fishermen of the 1920s (above) netted along the reefs of the atoll Ongtong Java
in a remote part of the Solomons. For special occasions, men on Ongtong Java would
gather as many green coconuts for drinking (below) as they could within a time
limit. The fruit would then be distributed among households.

A bite to snap the spinal cord was a simple means of dispatching
a fish before World War II and probably long before the arrival
of any Western influence. Subsistence fishing remains an important
source of protein, although stocks may decline if coral destruction
on surrounding reefs becomes severe enough to reduce habitat.

A glowering sky broods over the market at Gizo as do clouds of ethnic strife that broke out in late 1998 and continue to plague the islands, especially Guadalcanal and Malaita. Malaria and dengue fever are also rising. Even so, the mild climate and spectacular scuba diving opportunities continue to draw visitors.

the southwest Pacific. Scientists identify the Lapita as the original settlers of what we call Melanesia, plus parts of Micronesia and Polynesia.

The presence of Melanesian, Polynesian, and Micronesian influences in the Solomons is evidence of the early ethnic collisions and interminglings as these thousand islands were populated. The three distinctions themselves can be confusing to newcomers, in part because the terms have been misused for decades to describe ethnic types rather than geographic origin. Melanesia initially referred to the "dark" green mountain jungle islands stretching off New Guinea, but the term was misappropriated to mean dark-skinned peoples, who are roughly correlated with the "dark" islands. Melanesians are traditionally associated with New Guinea and the larger islands off its north and east coasts, whereas Micronesians come from the "micro" coral and sand islands to the northeast in the mid-Pacific, and Polynesians derive from the easterly volcanic islands such as Tahiti and Hawaii.

All three groups can be found in the modern-day Solomons, but the predominant group is Melanesian—the dark-skinned, wooly-haired group that more closely resembles Australian Aborigines or Africans—who linguistically and culturally have more affinity with the natives of Papua New Guinea than with any other islands.

The Lapita people were accomplished boatbuilders and sailors. This mobility allowed them to sweep ever southeastward, reaching the Solomons about 1600 B.C., Fiji and Polynesia about 1000 B.C., and Micronesia 400 years after that. Their trademark pottery contained geometric patterns, most often seen in bowls, pots, and beakers, and its remnants dot a huge swath of the Pacific from New Guinea to Samoa. Lapitans also dotted their path of exploration and settlement with fishhooks, pieces of obsidian, and beads and rings made of shells. The Lapita were primarily fishermen, but some evidence suggests they tried their hands at early agriculture and livestock cultivation.

What did they find in the dark-jungled Solomon Islands 3,600 years ago? They came upon a tropical paradise, much like what one finds today. The main islands were big and mountainous and forbidding, rising to 8,028 feet (2,447 meters) at Mount Makarakomburu on Guadalcanal. The Lapita paddled their canoes down the calm "lake in the sea" bordered by two parallel chains of islands, what the American GIs called the Slot. The western chain included what are now called Vella Lavella, Kolombangara, New Georgia, and Guadalcanal; the eastern, Choiseul Island, Santa Isabel, and Malaita. The two chains converged at Makira (San Cristobal) Island. The Lapita found a tropical oceanic climate, hot and humid, but with abundant rainfall. Temperatures hovered between 80° and 90° year-round, and 120 to 140 inches of rain fell on average per year.

Some of the Lapita explorers settled here. For those who stayed, over the centuries life settled into fairly predictable, if violent, patterns. Humans tended to cluster around small family-based villages led by chiefs known as "big men." These communities generally kept to themselves, except when shortages of staples, manpower, women, or other resources drove them to war. When they made war, they took few prisoners. The various island groups engaged in headhunting and cannibalism, and skull shrines from that warlike time can still be found.

As in Papua New Guinea, part of the legacy of tribal fragmentation was the fragmenting of language into dozens of tongues, each spoken by only a few hundred people. About 740 Papuan (non-Austronesian) languages extend from the Solomon Islands up through New Guinea to Timor. Linguists have been puzzled for years that the hundreds of languages spoken in this relatively small area seem to have no link to one another, except in the groups geographically adjacent. There is no discernible root language tying them all together.

Scattered eastward through the big islands all the way to Fiji are another 400 languages classified as Melanesian. Fijian is the dominant language in this group, but it is spoken by only 300,000 people. One Melanesian language prominent in the Solomons is Roviana, which was recorded and formalized by the Methodist missionaries, who recently celebrated their hundredth anniversary in the islands. Amazingly, the total number of people who speak one of the 400 Melanesian tongues is less than a million.

The island culture was Stone Age, because although its people were resourceful, they never learned the skill of beating metals into tools; indeed they were hard pressed to find any metals to work with, although some lay buried deep beneath the jungled peaks. These seagoing people practiced animist religions, worshiping deadly sharks that they regularly called to out of the waves, and looked to the spirits of their dead ancestors to guide them through the travails of life.

They expressed themselves in art, and the predominant color they used in their creations was black, with white and red used for detailing, and mother-of-pearl inlay for added style. They were fixated on the human head, real or imagined, and the head was usually the prominent part of their wood, clay, or stone sculpture. They created distorted and abstract figures that were generally squatting, sitting, or standing. Human heads were carved everywhere—on canoes, paddles, dance clubs, and the lodge poles of their thatched homes. Posts carved with more heads stood outside their homes, ready to receive the severed heads of their enemies. The warriors carried oval shields made of wickerwork into the battles where they might die a terrible death, or

In the 1920s, fishermen of Ongtong Java Atoll gathered in outrigger canoes to surround schools of fish and herd them toward shore for netting or spearing. Artificial lures made from turtle shell were also used, and it was reported that turtles were removed live from their shells and kept in pools to grow new ones.

Previous pages: The market at Gizo runs all day, goods packaged in woven palm leaves. To Max Kennedy, son of Robert, who came to witness the search for *PT 109*, watching the market operate "is like being in a biblical story. There's no refrigeration, but all the fish are sold each day. No one goes away wanting."

bring back a new supply of heads. On New Georgia, the more artistic of them built little shrines for ancestral skulls.

Their dugout canoes were also works of art, long and low with an upcurved prow and stern not unlike the warships of the Vikings thousands of miles away. The canoes were decorated with shells and perhaps the carving of a guardian spirit. Their art reflected their love and worship of the sea. It was the sea that had brought them to this place, the sea that brought them sustenance, and it was to the sea that many of them went when they died. Depending on local tribal *kastom* (custom), the dead were carried out to shallow reefs and left there to be taken by the sharks. The canoe houses where the men worked on their war canoes were supported by roof posts carved with the sacred bonito that they depended on for both physical and spiritual survival; there were also carvings of sharks and ancestors.

This civilization was musical, too. The Solomons were noted for their relatively sophisticated pan flute orchestras, capable of producing harmonies; this was unique in Melanesia, where the norm was a solo musician with instruments like blocks that were rubbed together, shark rattles (used to draw divine sharks out of the deep), crude flutes, and drums. The human voice was also an important means of expression, and that vocal tradition can be heard today in their Christian church services, where the strength of Solomon singers sounds almost unnatural to Westerners used to more restrained singing.

All of the physical markers of Solomon culture had at least some relationship to their religious and spiritual beliefs. Like much of Melanesia, the Solomons were shaped by beliefs in *tapu* (taboo)—certain forbidden or off-limits behaviors, places, and ideas—and *mana*, which roughly translates as "potency." While gods figured in myth, everyday life was infused with belief in the presence of dead ancestors and their importance in resolving present-day issues. The spirits of the dead were believed to temporarily inhabit animals like sharks, and while they were resident, the sharks were regarded as taboo.

In this primal world where spirits were observing and sometimes manipulating the physical world, the concept of taboo was supreme. Much like the Greek idea of hubris—excessive pride—that could bring on the wrath of the gods, breaking taboo would bring on the wrath of the spirits. Those who were foolish or courageous enough to break taboo were risking ruin.

The practices of Solomon cannibalism and headhunting have always inspired fascination and horror in Westerners, and in general, they have been misunderstood. Although cannibalism has occurred in many cultures around the world, those who engaged in it were motivated by different ideals. It has been alleged to have happened regularly in Native American culture, Aztec and Maya civilizations, in the Amazon and Africa, and themes of cannibalism crop up in European fairy tales as well as more modern sagas of shipwrecks, airplane crashes, and American settlers in the West. Although some have theorized that Mesoamerican natives engaged in cannibalism due to a lack of protein sources, there was no lack of protein in the Solomons then or now, as fish are abundant.

Instead, cannibalism in the Solomons, which may still have been occurring on an isolated basis when John Kennedy fought there—he wrote home about a native valet who had bragged about eating a Japanese—was an attempt to assert dominance over a foe. Eating a foe was a way to assimilate his mana. It may also have been a way to express contempt for one's enemy, as in the case of the valet that Kennedy heard.

Other bloody practices existed as well, such as lashing captured enemies side to side on a steep incline, and then letting a large sharp-keeled longboat slide down over them, dismembering them in the process. Islanders were also known to make sacrifices to appease angry spirits—in some case, hurling a girl to be eaten by the sharks.

THE INCA TOLD OF THEIR EMPEROR TUPAQ YUPANQUI sailing into what might have been these parts, and coming back laden with gold, silver, and black slaves. These legends inevitably made their way to the Spanish conquerors of Peru, never ones to turn away the prospect of finding mineral riches. A Spanish navigator in Peru, Pedro Sarmiento, spent seven years studying the Peruvian legends, and finally persuaded the crown to fund an expedition to look for the lost southern islands. But he was thwarted in his efforts to lead the expedition, having run afoul of the Inquisition, so that honor went to a very young sailor with better political connections, Álvaro de Mendaña de Neyra. He reached the Solomons in early 1568 under the flag of Spain; hence the Spanish names for islands like Guadalcanal, Santa Cruz, and Santa Isabel.

The first recorded contact between Europeans and islanders on Santa Isabel was almost idyllic, and too good to be true. The initial accounts by the Spanish purser told in adoring tones of Mendaña meeting with the chief, Bilebanara, and spoke of animated conversations about language and custom. Then Bilebanara, following kastom, offered to provision the hungry, weary Spanish. But when he realized the Spanish numbered 150, he must have changed his mind, because the provisions never materialized. A Spanish landing party was then sent ashore to peacefully but forcefully seek supplies.

Militiamen turned out on Florida Island in 1942 to train for defense against
Japanese invaders. Some seem to have rifles, others sticks shaped like rifles.
Both Japan and the Allies recruited local islanders to support their military
presence in the Solomons; many were used as scouts and spies.

This degenerated into a brawl and a battle, and from then on relations went downhill. The Spanish wrote of the darker, "unchristian" practices of the islanders, including cannibalism and animism, and relegated them to the status of brutes.

When Mendaña returned home to tell of his discoveries, his reports became more and more overshadowed by rumors about gold, embellished into the tantalizing speculation that these distant and long-lost islands held the fabled gem and gold mines of the biblical King Solomon. For that reason they were christened the Solomon Islands, a name sure to draw continued interest.

Despite the lure of gold, Mendaña wasn't able to make it back to his discovery for another 30 years, and during that voyage, rather than gold and riches, he found only disaster, dying of malaria not long after his arrival. The failure of his journey temporarily closed a door of contact with the West. As far as can be determined, the islands had no further European contact for another 200 years, although European navigators might have had glancing contact with some of the outer islands. Cartographers and explorers even began to question whether such an island group existed.

Not until the permanent English settlement in Australia did Europeans begin to sail regularly through these parts. Founded in 1788, Sydney was the hub of the new Australian colony. But that increased presence didn't translate into rapid European settlement of the Solomons, because the reputed violent nature of the island tribes continued to prove a deterrent. Anglican church workers transported islanders to New Zealand for education beginning in the 1850s, and they established permanent settlements two decades later. Catholic missionaries established a beachhead in 1898; Methodists founded a mission in 1902. At the same time, tens of thousands of islanders were taken to other Pacific nations, including Australia and Fiji, to work as plantation laborers, often in an exploitative manner known as "blackbirding."

An interesting side effect of the rising Western presence and Christian evangelism was the rise of cargo cults—religious movements that viewed the arrival of Western goods, whether on the wharf or on the beach from shipwrecks, as a divine act. Native anti-Westerners even began to claim that the goods were intended for the islanders but had been intercepted by the Western imperialists and must be reclaimed. Although these cults were stronger in Vanuatu and elsewhere, they also arose in the Solomons, and some continued even after World War II. One, the postwar Maasina Rule movement, exerted an influence on the country from 1944 to 1952. Although it was eventually suppressed, the nationalist impulse continued until self-rule was granted in 1975 and independence in 1978.

In tandem with Western religion had come European imperialism. Driven in part by concerns about exploitation of island laborers and also by growing French ambitions in the region, Britain in 1893 declared the islands a protectorate, and after World War I added the western Solomon Islands, including Choiseul, which had been under the control of Germany. A colonial capital was established at Tulaghi, and more British, Australian, and other settlers began to trickle in, largely engaged in colonial administration or the copra trade.

But while brutal traditions like headhunting and cannibalism diminished under the British colonials and the Christian missionaries, the new colonial rulers insisted on taxing their subjects for a government not of their choosing, and they also introduced common European diseases like measles, venereal disease, and influenza that drove the island population into decline. The islanders staged an uprising in 1927 on Malaita, when a British tax collection team was slaughtered. The act of protest evolved into a rebel movement led by one Basiana, who was captured in 1928. This was followed by his and other trials, Basiana's execution, which his children were forced to watch, and other punitive British measures. British rule survived, but an undercurrent of resentment toward the British continued.

The Japanese marched into this uneasy atmosphere in 1942. Those islanders who had benefited from the British and Christian presence sided with the Allies, but others felt no loyalty, or in some cases, tilted toward the Japanese, particularly in areas like Bougainville and Choiseul, controlled by Germany until World War I.

The Second World War was a baptism by fire for the Solomons, vaulting them at least for a few months into the global spotlight, and compelling the long-isolated islanders to confront an outside world that was richer, more powerful, and much more advanced technologically. Whether the Solomons are better off for having been dragged into the modern world depends on one's point of view. Most would argue that an end to headhunting and cannibalism and the introduction of modern civil traditions and modern health care are all advances. Others claim that the destruction of traditional institutions and the implantation of European institutions and ideas has been more harmful than constructive. A society that was once nearly classless now has an elite made up of government officials, businessmen, clergy, and others, perched atop a subsistence village culture that really doesn't participate in the modern economy or receive many of its benefits.

Politically, the eventual Allied victory ensured that the Solomons would enter the postwar era as a British territory, albeit restive with aspirations to independence and

Despite widespread conversion to Christianity, at least in its outward forms, many Solomon Islanders hold to ancient religious practices. At Laulasi Islet in Malaita Province, the spirit of the dead linger in the skulls of dead priests, since in this belief system the head is the dwelling place of the soul.

Previous pages: In a country with fewer than 25 miles of paved road but hundreds of islands, commerce goes by boat, as shown here in the harbor at Gizo. Most new craft are dugout canoes. Add an outboard motor, and this old technology works even faster.

redress of colonial grievances common to many developing countries. At war's end, the capital was moved from the ruins of Tulaghi to the former American base, anchorage, and airfield at Henderson Field on Guadalcanal to take advantage of the infrastructure that the three-year American occupation had created. The capital was renamed Honiara, and independence finally came in 1978.

Of its 480,000 inhabitants today, 93 percent are Melanesian, 4 percent Polynesian, and 1.5 percent Micronesian, the latter largely resettled from the Gilbert Islands by the British when their local water supplies began to run out. In addition, there are small numbers of Europeans and Chinese. About 70 vernacular languages are spoken, while pidgin English and English itself provide intertribal links. As in the past, most islanders reside in small, widely dispersed settlements along the coasts. Sixty percent live in villages with fewer than 200 persons, and only 11 percent reside in urban areas.

In one interesting development, because of Christian evangelism, the Solomons today are one of the most Christianized nations on Earth, with 95 percent following the faith. The islanders are unusually fervent and active in their faith. Affiliation implies more than a choice of identity; it implies churchgoing, activism, and proselytizing. Fire and brimstone street preachers are not uncommon in markets and public places, and the sounds of loud hymns roll through the towns on Sunday mornings.

Moreover, the Christian tradition has shaped other emerging institutions such as business and government. Many leaders and public figures come out of religious schooling, including a fair number who trained as clergy. While the church nourishes the soul and helps chart an overall national vision, the small island economy tries to nourish physical needs. Gross domestic product is about $900 million. The cash economy centers on natural resources like timber, fish, agricultural land, marine products, and gold. The main agricultural products are copra, cocoa, palm oil, palm kernels, and subsistence crops like yams, taro, bananas, and pineapple. Per capita income in the late 1990s was about $2,000 per year—not the poorest, but probably reflecting only those who participate in the money economy, and not those who live by subsistence and bartering.

Tourism, although small, was showing signs of growth until recently. Even so, those who choose to come are the hardy types, undeterred by long flying times, heat, humidity, the risk of malaria, and lack of creature comforts. Tourist attractions include reef diving and World War II wreck diving, trekking in virgin mountain rain forest, bird-watching, deep-sea fishing, untouched beaches, water sports, and native culture. Only a handful of international-standard restaurants and hotels are to be found, the domestic airline operates only a pair of small prop planes and one leased jet for a biweekly flight to Australia, and roads and medical facilities are rudimentary.

One precious national asset—and an area of critical concern—is the environment. According to the Nature Conservancy, the Solomons are among the ten most biologically diverse nations in danger of losing important ecosystems. The many islands possess hundreds of thousands of acres of virgin jungle, and the relative backwardness of the economy has served to protect many of the rain forests and their inhabitants, such as cockatoo, pigeons, eagles, ospreys, and other birds and small animals. But Asian traders have come in search of fish and timber, and the weak political institutions have a hard time resisting huge cash payments in exchange for their abundant natural resources. Some international groups such as the Nature Conservancy and World Wildlife Fund are struggling to raise environmental awareness, but it remains to be seen if the Solomons will avoid being overwhelmed by bigger global interests. Ecosystems still in existence but threatened if safeguards are not imposed include tropical alpine jungle, mangrove forests, grassland, dunes, wetlands, and coastal forest.

Makira Island hosts the densest population of saltwater crocodiles in the island chain, and these huge reptiles also pose a danger. Wherever humans live near their mangrove swamp habitats, attacks do occur. Most recently, an old woman on one of the islands near Gizo was attacked while in the village outhouse. The croc grabbed her by the head and attempted to drag her away. Miraculously she broke free, and although she required serious medical attention, she survived.

Other natural features include Marovo Lagoon, at 90 miles long considered one of the longest in the world. It connects a stretch of islands linked by reef and sandbars adjacent to the bigger islands of New Georgia, Vangunu, and Nggatokae. Mount Mbakararkombu on Guadalcanal is a challenging climb, and in the region is surpassed only by peaks on Papua New Guinea and in Irian Jaya. Kolombangara (6,000 feet/1,825 meters) is a two-day climb due to mud and jungle. These mountains and others above 5,000 feet (1,525 meters) in height are home to montane moss forests, abundant ferns, green algae, and a profusion of birds. Perhaps the most exotic denizen is the montane monkey-faced bat, although due to the remoteness of the region, other undiscovered species may inhabit the area.

While the natural environment faces challenges, the human landscape in the Solomons resembles Papua New Guinea and parts of Africa in its uneven transition to democracy and market economies. Honiara, with a population of 43,000, bears the closest resemblance to a modern city but other towns such as Gizo are little more than boat landings with a few basic businesses. The majority of people live in traditional

On July 7, 1988, with exuberant traditional dancing, men of Roviana Lagoon,
New Georgia, celebrate ten years of the Solomon Islands' independence, granted
by the United Kingdom. The disks and crescents suspended from their necks are
carved from fossilized giant clams.

family-based village compounds, and while lacking amenities like electricity, they seem tranquil and free of the tensions found in the towns, where unemployment and illiteracy are high (mid-1990s literacy was 64 percent). Many young men drift into the towns in search of work that does not exist, and conflicting tribal identities have led to political strife that in some cases has led to violence.

The most frequent ethnic disputes are between the natives of Guadalcanal and aggressive and determined immigrants from Malaita, who dominate the civil service and business community in Guadalcanal. This ongoing altercation resembles similar ethnic strife in Africa and elsewhere in the island Pacific.

At the other end of the chain, natives of Gizo and nearby islands have been drawn into the secessionist struggle of Bougainville islanders, who are fighting the central government of Papua New Guinea. This tension seems more likely to subside as the Papuan government concedes more autonomy to the rebels.

HONIARA IS THE ONLY PLACE THAT SHOWS AN URBAN BUSTLE. Young men are out jogging by the dozens before dawn, and as the sun rises through the huge tropical hardwoods and palms, open trucks carry men to various work sites around the capital. The downtown area is graced with a few modest high-rises. The market is large and busy, the harbor dotted both with arriving freight and passenger boats from the other islands and outside the country. Modern residences with iron gates rise up the forested hillsides outside town, and in British colonial tradition, schoolchildren in uniform march to school.

Gizo and the other towns, by contrast, languish in an unending torpor, broken only occasionally by the arrival of a passenger ship. The Gizo shoreside market is open 24 hours a day, selling fish and produce by day, and by candlelight selling meals to fishermen who head out onto the water. On the poorer streets, Gizo and the other towns are marked by scenes of tropical squalor—strewn garbage, an occasional dead animal, people sitting idly on the stoops of shacks.

The far-flung island villages seem almost unchanged by time, the only evidences of modernity being a cassette player or an outboard motor or a T-shirt. The villages and the family compounds seem calmer, more wholesome, healthier than the towns.

Into this faraway place of uneasy towns and tranquil jungle village we sail in May 2002. The past echoes here, including local lore about the American PT boat captained by a future President. It is a telling commentary of the islanders and their experience that they are not so much concerned with finding *PT 109*, but rather fear that, once found, *PT 109* will be salvaged and taken away from them. They are a people accustomed to loss.

152

THE DISCOVERY

CHAPTER FIVE

Loaded with electronic imaging gear developed by Robert Ballard and colleagues, the Australian vessel *Grayscout* begins sonar survey runs in Blackett Strait. First searches pick up too many targets to investigate. Using intuition, experience, and historical accounts, Ballard set course for the farthest southwest quadrant of the search zone.

MAY 2002

Thursday, May 16, 2002

I know that in this business, we will need some intuition, and some very good luck.
I would hate to come all this way with all these high expectations and be thwarted
by ocean currents, poor historical recordkeeping, or the passage of time. I would like
to say that all the gambling has been taken out of the process. But as the sun moves
behind a big tropical thunderhead to the west, I know that luck matters as much
as good calculations when you are trying to roll back the underwater curtain of time.
What of *PT 109* has likely survived on the ocean bottom from this long-ago collision,
and what would it look like to us 59 years later?

MONTHS AGO, WE HAD GONE THROUGH the preliminary negotiations needed to
uncover a wreck of this sort.

Communications flew back and forth between the National Geographic Society
and Caroline Kennedy Schlossberg and Senator Edward Kennedy. The family's initial
concern was that we might disturb the site and try to recover something. We reassured
that this was not our intention. Word finally came that Caroline Kennedy approved of
our venture, then that the senator also approved.

Cathy Offinger, our executive officer, went to work. She initially tried the folks in
Hawaii that we had used for the Pearl Harbor mini-sub project in November 2000, but

they were booked for May, the month we wanted to go. Then she contacted a dive operator in Gizo, to find a proper search vessel. He sent her information about boats in Guadalcanal, but none looked adequate for our needs. I told her about someone who had contacted me out of Papua New Guinea about providing us with ships in that remote area of the world.

A few days later Cathy called me, as usual out of breath. "Bob," she said, "I have a ship in Australia. It doesn't look that bad and the price seems reasonable. If we will make a firm commitment, they will hold it for late May." I agreed, digging a deeper hole in the funds I didn't yet have.

Not long after, we shipped most of our equipment needed for the dive. First it went to Texas, where our fiber-optic cable was placed on a new winch. James Stasny of Dynacon would loan us a brand-new winch ideally suited for the project.

Our two shipping vans went by truck from Texas to Los Angeles, where they were loaded aboard a containerized ship for shipment to Brisbane, Australia, and by truck from there to Gladstone, where the *Grayscout* was waiting. A last-minute air shipment brought all of our computers and other items our crew likes to keep close at hand until we are about to step on the airplane. On the operational side, everything was moving along flawlessly.

DALE RIDDER, MILITARY HISTORIAN AND MUNITIONS ANALYST, had taken a hard look at our situation. "I would be surprised if the hull is still intact," he wrote to me while we were still in the planning stage. "The watertight compartments were still holding but leaking somewhat when the crew abandoned. That gives three possible ways for the ship to have gone down: the watertight compartments gradually fill without collapsing, the compartment bulkheads collapse suddenly and create a water hammer effect which would explode the bow section, or the watertight compartments fail to fill quickly enough to compensate for the pressure increase and the hull implodes. Each mechanism will have some effect on the velocity of impact with the bottom and area covered by the debris field. The chances of the hull sinking in an intact state are about 50-50. My guess is that the hull remained afloat for an hour or two following the abandonment of it, and that it continued to drift southeast. If this is the case, it may have come to rest in or near the minefield laid by the U.S. in May of 1943, which accounted for three Japanese destroyers, which are also on the bottom. Add the odd Japanese landing barge or two, a missing U.S. WW II submarine (*Grampus*), and a fair number of aircraft in various states of disrepair, and there will be no lack of metal targets on the bottom. The fun part

In *Grayscout*'s control room, Max Kennedy, near left, nephew of John F. Kennedy, and Dick Keresey, in command of *PT 105* in August 1943, watch intently for images as Bob Ballard briefs them. Keresey recognizes a PT torpedo tube "within five minutes" by identifying a small hatch that opened for setting the torpedo's gyroscope. "I'd been looking at one every day for three years."

will be sorting them out. Oh, I forgot the 255 mine anchors and sets of mine cable of the minefield, and the possibility of encountering the odd mine that did not quite deploy as intended. I expect that the sonar guys are going to be very busy.

"The main recognition features of the wreck will be the weaponry, particularly the 37-mm gun lashed to the bow," he noted. "Another key will be the engines, which are a 12 cylinder Vee arrangement with mechanical supercharging. The radios should also have survived, at least the outer cases, as well as some of the galley equipment. The 37-mm gun, the .50-caliber machine guns, and the 20-mm are not excessively heavy. I would be wary of the torpedoes and torpedo tubes, in particular the detonators. I'll need to talk to Newport at length about those. The *109* may also have been carrying a couple of depth charges, used to discourage close pursuit by Japanese destroyers. If so, they will need to be approached with care. An underwater vehicle resembling torpedo recovery craft should have no problems with most of the gear. Concerning the crew's possessions, we might find some, assuming that we find the *109*. If they are organic material, they may not have survived too well."

Our work was cut out for us. But as if that weren't enough, a few days later another e-mail arrived from Ridder that gave us all pause.

"The U.S. laid a minefield in the Blackett Strait just after midnight on May 7, 1943. The destroyer-minelayers laid a total of 255 Mark 6 mines for the fun and enjoyment of the Japanese. The Japanese sailed three destroyers into the minefield the next evening. One destroyer managed to hit enough mines to sink immediately, while the other two hung around until the dawn, becoming target practice for 19 Marine dive bombers, who donated nineteen 1,000 bombs to the party. One destroyer took a bomb on the bow, and departed the surface for a rendezvous with the bottom, while the other was sufficiently annoyed by near misses to do likewise.

"The problem is this. It was not until I compared the map of the minelaying in Morison's *Breaking the Bismarck's Barrier* with the overall layout map in *Marines in the Central Solomons* that I realized that the Morison map was incorrect as to the location of due north, which was the course of the mine layers. I also realized that if the minelaying was done according to the map in Morison, the last half of the pattern would have been laid somewhere in the middle of Kolombangara. As ships sail very poorly over land, I then got serious with a ruler. The minefield was laid in three lines, each line consisting of 85 mines laid 100 yards apart, with the minelayers approximately 150 yards apart. This would make the field 8,400 yards long and 300 yards wide. When I adjusted for the incorrect true north, and also compensated for the near proximity of Kolombangara, the following likely location of the minefield emerged. It should run from just north and west of the north entrance of Ferguson Passage for 8,400 yards or roughly 4 nautical miles. I need to lay this out exactly on a map, but it puts the minefield in the middle of the most probable search area for the *109*. I also did some probability crunching on my calculator. If you assume 99 percent reliability for the mines to correctly deploy, there is less than an 8 percent chance that ALL of the mines deployed properly. If you assume a 99.9 percent deployment reliability, the chance of correct deployment of all 255 mines is 77.5 percent. Given this, I very reluctantly conclude that there is a high probability of at least one mine being encountered somewhere between the bottom and say 50 feet of the surface. The Mark 6 mine was designed in WW One as an antisubmarine mine, with a 300 charge of TNT. Its detonating agent was a copper antenna wire, which generates an electric current on the near, like about a foot or so, approach of the metal hull of a ship. It cannot be disarmed, and had no provision for self-destruction or sanitizing.

"Before putting anything major over the side of the *Grayscout*, we are going to have to do a sonar sweep for any targets located above the bottom. I hope that this does not put a major damper on things. Since I know how the mine works, I know how to work around it if we are so unfortunate as to find one where we do not want it to be. Also, the location of the wrecks of the Japanese destroyers on the bottom should give us a much better idea as to exactly where the minefield was laid."

Dwight Coleman, who runs navigation and sonar for my team, noted that towed side-scan sonar would have two frequencies, relating to two resolutions. The resolution would be dependent on towing altitude off the bottom and range and speed of the vessel. Highest resolution would come from towing at about 1-knot speed (a little more than one mph), 150-foot range, and 15- to 30-foot altitude. This should resolve features on the seafloor about the size of a cinder block. He thought a mine should show up at a coarser resolution, although you wouldn't necessarily know it was a mine. The coarsest resolution would be to tow at about 4 to 5 knots, 1,500-foot range, and about 250-foot altitude. For this, the smallest resolved feature would be about the size of a pallet of cinder blocks.

Dwight and I also knew that "beam spreading" also affects resolution; objects at the far ranges are not as well resolved as objects in the near ranges, for the same swath and speed setting. Optimally, in search mode, we will probably run at 3 knots, 150 feet altitude, and at a 1,000-foot range. At this rate, we can cover about 10 square miles in 10 hours. Obviously there will be a trade-off between resolution and area coverage.

With a target identified, the crew on *Grayscout* make ready to launch *Argus*.
She will provide lighting and backup video from a depth of nearly 1,300 feet. *Argus,*
tethered to the mothership by an umbilical cable, can be piloted to the desired
position to hover and maneuver by panning and tilting.

Pages 158-159: A Solomon Islander paddles a dugout past tiny Kennedy Island
(then Plum Pudding) where the crew of *PT 109* struggled ashore. After four
days they swam on to Olasana Island. Max Kennedy swam that same passage 59
years later, "a long, tough swim but not as tough as doing it at night with enemy
all around. I'm glad I did it and glad that I will never do it again."

Dwight and I asked Dale if he had rough coordinates for anything he knew about on the seafloor in Blackett Strait. We could design a decent search box for the portion of *PT 109* that sank, but who knows where the floating section ended up?

Perhaps closer to Ferguson Passage. We could map the entire Strait and passages in a few days, which would give us plenty of targets to inspect. The trick would be to positively identify the targets with the ROV.

A later e-mail from Dale went into more detail about what of *PT 109* he thought we would find on the bottom of Blackett Strait: "As I indicated, based on the loss report and on who survived, the boat was not cut in half, but had the starboard side sheared off. This is about the only way to account for Kennedy surviving but the starboard machine gunner being killed."

Dale theorized that we would find a number of items located near the sunken hull or two pieces of hull, including torpedo tubes, engine or engines, one .50-caliber twin machine-gun mount with metal bars around it, one 37-mm antitank gun (the identification key), one 20-mm automatic cannon with mount with no shield, two Packard V-12 supercharged engines with shafting and propellers, one smoke generator tank, the radio equipment, and the galley equipment. Because the deck was made of mahogany planking, with the internal bulkheads fabricated of laminated spruce and white oak formers covered with marine plywood, he thought we should check out the ability of white oak to survive under marine conditions for 60 years.

Dale also questioned how to interpret Coastwatcher Evans's statements that he might have seen the hull floating as late as August 5. Dale doubted that it was the *PT 109* after August 2. According to the report of loss, the crew abandoned the hull when it turned over and was awash. At that point, the hull probably stayed afloat at most for four more hours. He also pointed out the statement in the Tregaski book that the hull was abandoned at ten in the morning, not two in the afternoon, which makes much more sense. If the starboard side was sheared off, the stern and midships section of the hull would have been providing little if any flotation. The torpedo tubes add 9,000 pounds in weight, the engines and shafting about 7,000 pounds, the 20-mm about 1,500 pounds, and the twin .50 mount about 750. Add in the smoke generator, and you have well over 18,000 pounds of weight in the stern section pulling the boat down. You also very likely have a small but continuous leak of aviation gasoline from the submerged fuel tanks. He theorized that the crew constantly smelled avgas fumes, and were afraid of igniting them in any fashion. Any PT crew had a great appreciation of the explosive properties of avgas.

163

When I pressed the point that several sources including Evans, the Australian Coastwatcher who received the coconut shell message from Kennedy, talk about the hull of a small boat seen floating until August 5, even running aground, Dale quoted from Evans's August 4 dispatch in Walter Lord's *Lonely Vigil*. "Cannot confirm that object seen was floating hulk of PT. Object last seen approximately two miles NE Bambanga drifting south, not seen since PM Second." Evans did not see the "object" on the 3rd or 4th. On the 5th, something was sighted in Ferguson Passage, drifting south. Based on where Evans was positioned, it seemed unlikely that something could be visible on the 2nd, disappear for two days, and then suddenly reappear on the 5th. Most likely the "forepart of a small vessel" was wreckage from a Japanese landing barge or small craft, not the wreckage of the *PT 109*.

MY TEAM WAS SUPPOSED TO ALREADY HAVE COMPLETED two days of sonar runs in Blackett Strait by the time I arrived. Instead I find out I beat them to Gizo by a few hours; they haven't even started yet. In addition, *Grayscout* is two days late in arriving from across the Coral Sea into Gizo Harbor. That forces us to adjust our schedule further. On top of all that, I learn that a critical container of our search equipment was inexplicably delayed in getting to Gladstone, Australia. And sailing time from there was about a day longer than our skipper had predicted.

Other than sitting in the few air-conditioned spots in the country, or floating in the ocean or a swimming pool, we are forced to sweat it out, literally. Cloudy or sunny, day or night, the often windless air hangs heavy with humidity. The thermometer dips no lower than the 70s at night, and most days climbs above 90. Coming from the chilly New England spring, I feel the tropic heat like hot breath all over me.

About noon, those of us onshore can finally see *Grayscout* coming into view in Gizo Harbor. Some of us cheer, thinking we're about to get started on our search. But the ship can't dock, because none of the customs and immigration officials can be found to receive it. Needless to say, we cannot begin our work without the proper government approvals. So it idles just offshore, tantalizingly near, while we wait.

The government inspection is finished in time for us to take a late afternoon ride into Blackett Strait. In the east, Kolombangara looms above all, sheathed at several altitudes by gray wisps of afternoon cloud, some bearing rain, all turning gold and rose as the sun slowly sinks in the west. In one vast sweep I eye the circle of islands and peaks all around me—Gizo to the south and west, Vella Lavella far off to the northwest, and to the southeast, the narrows between Kolombangara and Kohinggo Island where the *Amagiri* and the other Japanese destroyers raced out that night en route to and from Vila Plantation. Farther southeast I can make out New Georgia and Rendova, the latter where Kennedy had his base at Lumbari.

The ocean is like indigo glass today, hardly turned by current or wind. In this beautiful sunset, it's hard to imagine that a war was ever fought here. At last I look away from the panorama. Time is of the essence, for *Grayscout* must return to her home port on May 26. That gives us seven days—seven days to find John Kennedy's lost boat, the location of which we only vaguely know. We lower *Echo,* our side-scan sonar that will begin "mowing the lawn" of Blackett Strait—towing the sonar vehicle up and down the length of the Strait to make a sonar map of the bottom.

Saturday, May 18
We mapped sonar lines all night. When we prepared to launch *Echo* for the second day, we found a problem with the signal coming back from the "fish." It took some five to six hours to track down the problem—which turned out to be human error. Dave Wright had unknowingly put an optical inhibitor in the junction box and so lessened the signal that the software couldn't detect it. With the problem solved, *Echo* was launched around 2200.

Glitches are to be expected. For one thing, the crew and skippers of *Grayscout* are not accustomed to scientific expedition and have to learn our techniques on the spot. Key among these is holding *Grayscout* steady in whatever position we ask, despite the effects of wind, current, and balky engines or controls. Although holding a straight line for a distance of five miles without deviation is impossible without dynamic positioning thrusters, which we don't have on this boat, we strive to get as close as we can.

One added inconvenience is that our impromptu control room on the *Grayscout*—home to the many computers, LCD displays, and video monitors, plus keyboards and other electronic controls that show the results of our sonar and video scans of the bottom—is proving to be even hotter than it is outside, at a perpetually miserable 94°. Neither fans nor air conditioning are having any impact.

Our search area is approximately 35 square miles, a grid of seven by five miles, between the islands of Olasana on the southeast and Kolombangara on the northeast and proceeding northwest up the strait in boxlike fashion. We have had considerable discussion about the likely position of *PT 109* when she was hit, and when she sank.

Popular wisdom holds that the boat was hit in the middle of the strait. But Kennedy and his men and the crew of the two other PT boats nearby were all uncertain of their

position when they were hit, having become separated from their command boats. The crash with *Amagiri* came at 2 a.m., and according to the survivors, the larger bow section remained afloat, while the smaller stern, weighted down by engines, went straight to the bottom.

But some, including munitions analyst Ridder, who is along to help us identify *PT 109* and other wrecks by their armament, theorizes that instead of breaking cleanly in two, the collision instead shaved a narrow slice off the *109* hull that sank immediately. The rest of the hull floated because the watertight bulkhead held for a time.

But how long did the bow float, and how far? We gradually eliminate wind as a factor, not because there wasn't wind, but because the hull was so low in the water that wind had less to push against. That leaves current. How does the current flow?

Our local contact tells me the current tends to flow north through narrow Ferguson Passage for part of the day, and then reverse itself later with the tides. That suggests that as long as *PT 109* was afloat, it drifted back and forth in the Strait. While the hull might have floated into the more turbulent waters of Ferguson Passage and stayed there, my team and I conclude that the odds are that the hull stayed in Blackett Strait rather than flushing on into Ferguson Passage. Part of that may be wishful thinking—it's easier to scan this calm sector of water than to fight the currents and tides in the passage. In the end, our consensus is to first look for the boat remnants in the area closest to where *PT 109* was hit in Blackett Strait before we go wandering afield.

We plow up and down the strait, *Echo* doing her work, sending real-time sonar scans up to our monitors. Our onboard control room is a hot, dark, and crowded part of the aft cabin that has been stuffed with a half dozen flat-panel display monitors showing the pattern of the sonar scans, the position of the ship, readouts on the bottom depth, our speed, the water temperature, and the depth where *Echo* is suspended. Behind and underneath the monitors are portable modules of electronic and computer equipment that support these displays, plus three magnificent high-definition television monitors that show pictures when the other vehicles are down.

The sea depth rises and falls from as shallow as 800 feet to as deep as 1,500 feet as we plot out the entire five-by-seven-mile grid. The *Echo* can alter our scanning distance from about 1,000 feet to as much as 1,600 on either side, for a total scan of more than 3,000 feet. Of course, there's the gap of bottom directly below us that we miss.

Jack Kennedy's PT boat was less than 90 feet long to begin with. In all likelihood, worms have eaten away most of the mahogany and plywood. But even so, some parts would still remain on the bottom—the machine guns, engines, and torpedo tubes were metal instruments that should ring loud and clear on sonar.

It should show up unless we're in a very rugged area. Parts of the search area are volcanic, so that means metal is going to be hard to pick out. We'd love the wreckage to be located on a very muddy bottom, where it would stand out. We're hoping we get lucky.

Grayscout tows *Echo* at a heighth of 150 feet off the bottom. It bounces sound waves off the rocks and coral and the mud, and paints an electronic picture of the bottom. It's a tedious business, and almost hypnotic to watch the blurry sonar squiggles as they scroll up on a computer screen.

We have no lack of sightings. At 10 in the morning we pick up what looks like a Japanese landing barge. At 10:30 we see a profile very much like what a Japanese destroyer would create. At 11 we suffer a few glitches with our sonar itself, although it seems to be with the real-time display monitor and not with the data that are being recorded. At 11:30 we see another possible landing barge, or a pile of rocks.

As we ride along, an earlier question of mine—why the U.S. depended on PT boats to fight such a powerful force of destroyers is answered.

In this lake in the ocean, I can see why the U.S. didn't want its destroyers trapped in this giant bathtub. It's too small to safely and easily turn about to face an opposition force. There's no maneuvering room, and aside from handicapping actions against the enemy, it would have compounded the incidents of friendly fire. Problems with misidentification were constant, and the results often fatal.

Another factor, which strikes me forcefully during our 24-hour operations here in Blackett Strait, is how thoroughly night descends on this series of lush tropical islands. When the sun goes down you feel like you are entering a pitch-black box.

As the day wraps up, I feel good. Although we're doing 24-hour shifts and things aren't perfect, they are well within plan. We'll map this area of the Blackett Strait, examine our possible targets by video, and then, if necessary, take the boat and crew down into the more challenging Ferguson Passage, where currents can easily knock us off-line. I want *Grayscout* to be more comfortable with my methods before I do that.

Sunday, May 19
Our sonar slammed into the side of a volcanic vent never before detected, which seems to jut out from the side of Kolombangara. If active, it's quite a discovery. The water temperature in that area is a few degrees warmer than the surrounding ocean.

In any case, that incident necessitates a six-hour repair job, reconnecting severed fiber-optic cables. The clump weight took the brunt of it, causing a dent in its lead and steel nose; *Echo* got a ding in its syntactic nose. The cable on the clump weight was also cut.

Just after dawn I confer with my crew. We'll finish up Blackett Strait on Sunday, then do a trial run down into Ferguson Passage just to see how we perform. Our captain, who has radar and modern charts, has asked us not to work on the edges of our search area even though we assure him we know exactly the depth at which *Echo* is working. Crashing it into the bottom last night didn't help build his confidence.

We are in the middle of our search area, looking in the southwest quadrant for the bow of *PT 109*. My strategy is to concentrate over the next 24 hours on the northeast and southwest quadrants, and then to deploy *Argus* and *Little Hercules* for the first time tomorrow morning.

BY 8:30 WE COMPLETE OUR SONAR RUNS on the Blackett Strait box that we had laid out. It's time to pull in *Echo* and put out our video scans, *Argus* and *Little Herc*. These two space-age vehicles are critical to my undersea searches, and they have traveled all the way from Mystic, Connecticut, to help us. The larger vehicle, *Argus*, is a steel-gray, squarish fiberglass pod about the size of a living room sofa. It has powerful underwater lights and its own video cameras intended to monitor *Little Herc*. That craft is smaller and bright yellow with metal rods and fittings. It moves independently, yet always remains connected to *Argus,* which is in turn attached by cable to the ship's winch. Although *Echo* can produce only a printed pattern of bottom features that looks much like an electrocardiogram you get at the doctor's office, *Argus* and *Little Herc* will produce real, commercial-quality video pictures.

The effort to bring in *Echo* goes as planned—I help out on deck, together with Mark DeRoche, Cathy Offinger, and Dwight Coleman. At one point the sonar rig swings wildly, but quickly four sets of hands reach out to steady the precious equipment.

Dwight and I do a complete review of all of the sonar records and associated targets. The results are, to put it bluntly, depressing. There must be several hundred if not thousands of targets that could be a piece of a PT boat. Looking at all those targets, I'm beginning to wonder if finding the severed stern of *PT 109* is in fact possible.

I spend some time poring over my naval charts and records, and talking to Dale Ridder. If I was expecting to have found *PT 109* by now, I would be disappointed. But I know that we are attempting something very difficult—finding a small wooden vessel in an area affected by tidal currents, volcanism, wood borers, and the passage of time.

An increasingly worrisome question is: Do we have enough time to do the job?

On Kauvi Island, Max Kennedy talks with Eroni Kumana, center, and Biuku Gasa, the two scouts who first encountered the marooned crew of *PT 109* and later took the message coconut to Rendova. For Kennedy, "it was a very emotional and moving experience. Many things were spoken without words."

Pages 166-167: Following *Argus,* and tethered to her by cable, *Little Hercules* goes over the rail, guided by Bob Ballard, at left. *Little Herc* has a greater range of action than *Argus* and has been equipped with an HDTV camera to give exceptionally detailed images of the target.

Tuesday, May 21

At 0800, we launch the vehicles. The sunrise is beautiful and the seas a flat calm, an inviting lure from King Neptune. The launch is flawless, as is the trip down toward the bottom. Yet when we hit a depth of 600 feet, the system begins to crash. *Little Herc* is losing power and *Argus* is having thruster problems. We reset them, but they crash again.

We bring the vehicles back up to the surface but do not recover them, slowly towing them in the water, troubleshooting. The burden falls on team members Jim Newman, who built these two vehicles, and electrical engineer Dave Wright.

A great deal of the trouble we've been having stems from the tremendous heat we've been forced to work in. High temperatures inside the pressure housings of the vehicles, the power system on the surface, and within the control space aboard ship are all taking their toll. Since I can't do much to help my technicians other than stay out of the way, I end up grinding my teeth in frustration until we can get our systems operational.

IN THE AFTERNOON, WE STOP AT THE HOME OF BIUKU GASA, one of the two island boys who found John F. Kennedy and his crew in August 1943. The family compound is nearly the size of a small village of thatched huts on stilts, surrounded by emerald green grass, small family plots planted with fruits and vegetables, a few chickens and dogs, and a handful of great-grandchildren. Biuku receives me in what passes for a parlor in these parts: an open-sided thatched shelter. He is almost 80 now, and prefers to remain seated when I greet him. The family acknowledges my expedition by cutting down a thick tree for carving into a commemorative dugout. I'm honored by the gesture.

As the visit ends, I see that the sun is getting lower in the sky. We need to make the 30-mile return trip to *Grayscout* before darkness falls. We speed past mile upon mile of what looks like virgin South Pacific jungle, broken here and there by a hut or family compound; the only traffic is an occasional dugout. The whole scene is tinged gold by the declining sun. We reach the research ship just at dusk, after a glorious display of sunset pinks and golds in the west.

The video and submersible glitches have been resolved, and we are back in business. Entering the control room, I'm once again impressed at the clarity of the video images we are getting from below. While I do not spy a PT boat, I'm seeing tropical fish, schools of shrimp, blocks of coral, volcanic rock, and even individual grains of sand as clearly as if in a brightly lit home aquarium. It is this futuristic television-like apparatus that may enable us to find *PT 109* in the middle of a big and mysterious ocean.

Wednesday, May 22

After four days of looking for the smaller starboard slice of *PT 109*, I've changed my approach. Last night I decided instead to look for the larger bow piece of *PT 109* that drifted toward the Ferguson Passage for at least half a day after the collision.

Our time constraints forced me to take a gamble. We initially chose to look for the stern because that section sank instantly and we had a better feel for where Kennedy was hit than where the bow might have drifted.

But in looking for the stern, we were focusing on a search area that included hundreds of targets, most of which were rock outcroppings, junk, and other objects. Going from target to target with our two vehicles would be slow and tedious, and might have required another week. We didn't have a week. *Grayscout* needed to return on May 26.

In deciding to look for the bow, I knew I was looking for something that had drifted that resembled a hull in shape, was up to 80 feet long, and was probably isolated or anomalous to the sea bottom in that area. Luckily, the bottom there was almost devoid of distracting clutter, unlike the area where the collision had occurred—perhaps because the currents there had either swept away foreign objects or buried them in sand. By moving to the new search area, I was cutting to the chase.

Back in early August 1943, after the PT base at Rendova realized that *PT 109* had sunk and concluded that the crew hadn't survived, they focused on the floating bow that Coastwatcher Evans reported seeing floating south down Blackett Strait toward Ferguson Passage. Rendova had requested that Evans work with his scouts to ensure that the supposed bow section was destroyed. To assist in this effort, Evans decided to leave his outpost on Kolombangara overlooking the strait and move to a small island near the Blackett Strait side of the entrance to Ferguson Passage. When he reestablished himself near Ferguson Passage, he reported seeing floating wreckage that was hung up on the reef outside the passage near Naru Island.

New Zealand P-40s strafed this wreckage, leading Kennedy to wrongly conclude that Naru Island must be occupied by Japanese; he thought it was the logical place for them to be to monitor U.S. boat activity in and out of Ferguson Passage.

Native scouts Biuku and Eroni also saw the wreckage hung up on the reef near Naru and went over to it and removed two Japanese rifles, which they told me when I met with them. The object hung up on the reef at Naru was a Japanese boat or barge, not *PT 109*.

The floating *109* bow section must have sunk sometime after Kennedy and his crew left it. Because of the currents in that area, I concluded that it must therefore be within the Blackett Strait. With this thought in mind, I had Dwight Coleman extend our

As seen from *Argus*, *Little Herc* is positioned over the target (above), then comes in for a closer view of its own (opposite, above). The object in the foreground is a scraper attached to *Little Herc* by two struts. Beyond is a torpedo tube (opposite, below) from *PT 109*, seen in another view.

The torpedo tube with the front part of its torpedo still inside, right, and the after-body, left, rest quietly on the bottom, and will continue to do so. Two men died on *PT 109* on that August night, and the remains of the vessel are being treated as a grave site not to be disturbed.

sonar coverage in the southwestern quadrant as close as possible to the islands at the southern end of the strait. I also had him run our westernmost lines as close to Naru and Kennedy (the renamed Plum Pudding Island, where Kennedy first landed) as possible. The flanking coral wall in that area is so steep our ship could get within 100 yards of the reef—and since our sonar could see 1,000 feet to either side, we would be able to view everything in deep water.

The survey was completed, and Dwight and I reviewed the sonar targets. The results were stunning. Unlike the northeastern quadrant of our search area, which contained hundreds to thousands of targets that might represent the sliver shaved from *PT 109,* the southwestern quadrant contained only one target!

Surrounded by a featureless sea bottom, the target was long and narrow, shaped like a ship's hull. It was rectangular in shape measuring 30 to 40 feet by 23 feet—just about the width of *PT 109.* It also contained a series of strong targets that looked like torpedo launchers.

I took a deep breath and sent *Argus* and *Little Herc* in to take a closer look.

We found what at first looked to the untrained eye like a section of rusted, encrusted cylinder or pipe, maybe two feet in diameter and about thirteen feet long, but broken in two halfway down. Anenomes waved their tendrils from its barrel. Fish swam in and out. This otherwise unexciting cylindrical object lay in a small depression surrounded by gently sloping walls of sand on a remarkably clean bottom that looked like the Sahara.

The tubes looked like either torpedoes or torpedo launchers or both. We would have to wait until our historian Dale Ridder and *PT 105* skipper Dick Keresey had a look.

We ran our video cameras from several angles, then shut down for the night. Though my little bunk on *Grayscout* was no more comfortable than it had been the nights before, I slept better than any time since I'd left the U.S. days ago. As I drifted off, I was almost pinching myself to make sure that my last-minute discovery wasn't just a pleasant dream. While I would wait for the experts' final verdict, I felt deep down that we had found what was left of *PT 109.*

THE NEXT MORNING OUR SHORE-BASED CREW came out. Joining them, after just arriving from a long trip from the U.S., was Dick Keresey. He had fought alongside Kennedy, had been searching in Blackett Strait that fateful night, and he knew PT boats better than nearly anyone. Though jet-lagged after a series of flights from Florida, he was ready to be of any help he could.

With him was John Kennedy's nephew, Max Kennedy, second son of Robert F. Kennedy. He had flown all the way from Cambridge, Massachusetts, where he and his wife teach literature courses at Boston College. Amazingly lucid after a 38-hour trip across ten time zones, Max too sat down in our control room, and we ran the tape.

We went back over the object from different angles, stopping, rewinding, zooming in. The silence was heavy, broken only by occasional commands to the video crew.

Again we saw the rusted cylindrical object, man-made but looking this time more like a natural feature of the bottom. Some of us speculated out loud about a torpedo tube, whether we could see the remnants of a torpedo inside, and why the object looked broken midway down its length.

Both Dick Keresey and Dale Ridder ventured that it looked like a torpedo tube. We could also see the corroded propellers of a torpedo at the rear of the tube. Dale thought he could make out the hemispherical object that was part of the apparatus to launch the torpedoes out of the tube. And he thought he could detect the strap that anchored the tube to the deck.

Keresey felt a sense of instant recognition. "I looked at torpedoes just like that for 18 months of my life," he said. "I was convinced that I was looking at the port torpedo tube and a torpedo from an American PT boat." Dale agreed, and ran to his voluminous portable library of histories, catalogues, directories, and other guides to American and Japanese vessels. He wanted to confirm the exact size of the original torpedo tubes on a boat like *PT 109*, find any information about the mounting straps and the pressure mechanism that ejected the torpedo from its tube when firing, and also ascertain whether any other PT boats or PT torpedo tubes had been lost in Blackett Strait during or after World War II.

By one o'clock, I got as much of a confirmation as I would be able to get in this remote corner of the world. Dale came back from his portable library in the galley into the control room, nearly shouting in his excitement.

"This is it!" he said. "This is the port torpedo tube from *PT 109*. Everything matches. There were no other PT boats sunk in this area. I have the record of the disposition of every PT boat ever built. Only *PT 109* went down here. The torpedo tube matches what we would be expecting to see. The strap matches. The size matches. This is it."

The supposed torpedo tube from *PT 109* was resting in 1,300 feet of water. The ambient ocean temperature around it was about 11° C, or 50° F. The hemispherical object at the rear of the forward section of the tube was the air flask that helped launch the torpedo. At the time of manufacture, it was positioned at the weakest part of the tube, so it was logical that the tube would break there. We also had confirmed that this was a port side torpedo tube; we just couldn't tell if it was fore or aft.

The hull was pretty close to where Ridder, Dwight Coleman, and I had calculated it would be. Mercifully, we hadn't had to negotiate the U.S. minefield Dale had warned us about early on. The Japanese may have swept the field, even though there was no record of them doing so. All we knew was that there weren't mines anymore.

Now that we had found what was almost certainly a piece of *PT 109*, I was asked if there would be any recovery effort. "That would be up to the Navy," I answered. "This is considered a grave site."

But my thoughts were running in a different direction. The huge mound of sand around the tube led me to believe that the rest of the hull was buried there. The current runs very strongly in that sector; any object on the bottom acts much like a snow fence on land. The current is so strong that all the silt has been scoured out of the heavier sand, much like a winnowing. That makes for much clearer water. The sand piles up around the object and then covers it. It may get uncovered every few years by storms, changes in current, and so on. Considering those variables, it was miraculous that we had found anything at all.

Our sonar image revealed a larger target than we were seeing with the video cameras. That's because the sonar signal can penetrate into the bottom. It showed an object measuring 23 by 40 feet, including several hard targets. Twenty-three feet just happened to be the width of *PT 109,* and I was thinking that the other hard targets were the buried engines. Although I hadn't come out here to do any bottom excavating, my curiosity was piqued. I wanted to know what was underneath. Was the mahogany deck and plywood hull intact? Mahogany buried in the sand would last a very long time. Would we find the cockpit or the ship's log or something that clearly established the ship's identity?

Knowing it was a long shot, Max Kennedy authorized me to send an e-mail back to Senator Edward Kennedy that night, asking his permission to scrape some sand away with a probe to see if we could hit deck beneath. It was after midnight on the U.S. East Coast, so I knew I'd get no reply until the next morning.

The sun was setting in the west. Whereas 24 hours ago I'd been concerned that we were getting nowhere fast, now I felt a sense of relief. We had scored. Every bit of information available to us said that we had found part of John F. Kennedy's *PT 109.* We had brought a lost chapter of history to life.

Back in Washington, three experts review tapes of Ballard's discovery. Welford
West, torpedoman aboard *PT 157,* Mark Wertheimer, Assistant Curator at
the Naval Historical Center, and Claire Peachey, underwater archaeologist, concur:
The image is of a portside torpedo tube from a PT boat.

Pages 184-185: Max Kennedy, a young nephew, and Dick Keresey, an old comrade,
walk the beach of Kennedy Island. Kennedy recalls Keresey's brave action on that
long-ago night. Keresey recalls orders that let radar-equipped PTs return to base
and leave those without blind in the strait: "I couldn't understand it then or now."

Friday, May 24

Our run of dry weather is at an end. The day has dawned blustery and dark with rain squalls and wind giving the ocean a strong chop and swell and turning the dazzling green Solomon Islands a foreboding gray.

Despite the dark weather, spirits are high on *Grayscout*. With "permission to scrape" in hand from Senator Kennedy's office, we send *Argus* and *Little Herc* back down. This time *Little Herc* has a small arm composed of a piece of steel wrapped in tape. This is all we can improvise on such short notice.

After a number of maneuvers, we return to the broken torpedo tube and curious sand formation that tells me a PT boat lies beneath. After a tantalizing wait as the vehicles descend, the screen shows once again the torpedo tube. Before we start scraping, we take a good long look at what we've found. The tube and torpedo, broken in half, are slightly separated and lying at an angle to each other. When viewed from the front of the tube, the retaining ring for the protective cover, which would have been removed prior to leaving base, is clearly visible, lying on the bottom directly below the front of the tube. The front portion of the tube is split somewhat along the weld line. The retaining straps are clearly visible and correctly located on the front portion. They are still attached to what appears to be wooden blocks, which may still be attached to the deck.

Visible as well is the housing of the gear used to crank the torpedo tube outboard prior to firing, with the position of the gear showing that we are looking at a tube that would have been located on the port side. Farther aft, at the point where the torpedo had broken in half, the aft end of the torpedo air flask is clearly visible. The weakest point on the torpedo was immediately to the rear of the air flask, which was designed to hold air under 2,800 pounds of pressure. So breaking at that point is perfectly consistent with the characteristics of the torpedo. The opening for setting the torpedo gyroscopes, which maintained the torpedo on a correct course, are clearly visible and correctly located at the front of the rear section.

Looking up inside the torpedo and tube, the housing for the gyroscope can still be made out. The retaining straps for the rear section on the tube are still present, although quite rusted, and on both sections the reinforcing rings around the tube are faintly visible. The rear of the tube is rusted away, and the fins and contra-rotating propellers are quite visible. In comparing the tube and torpedo on the bottom with the picture of the tube and torpedo from a reference photo, they are clearly those from an 80-foot Elco-class torpedo boat: *PT 109*.

Our scraping maneuver is easier said than done, considering our ship doesn't have dynamic positioning, an automatic system using satellite positioning and thrusters to keep a ship exactly set in place. We must do this manually, which means Dwight Coleman must continuously radio steering commands from the control room up to the skipper on the bridge. When the ship and *Little Herc* inevitably drift out of position, it takes us some time to get them back on the dime. Finally we are positioned correctly, and we start scraping to find the deck. Unfortunately, the process begins haltingly. Soon it resembles the arcade game in which you drop the claw to grab watches and cameras and you end up dropping the claw onto nothing.

We try to pull the sand. We try to push. And while we might get a clump to move, on our backstroke we knock it back to where it was. We stir up bottom sediment that blocks our view. We drift off position. We keep trying, but nothing happens.

After an hour of this, I see that it's hopeless. There's too much sand, and our tools just aren't that good. But before giving up, we try one last maneuver. Easing over to what we think is the cranking mechanism used to launch the torpedo, we try to move it. If it is just a small piece of metal sitting on the sand, moving it should be easy. Pushing as hard as we can, it doesn't budge. It is clearly attached to something just below the sand. Perhaps it is the deck and hull of *PT 109*. Everyone feels some disappointment as we wind in the winch, bringing *Argus* and *Little Herc* back to the surface for the last time.

But the longer I reflect on this disappointment, the more appropriate it seems that the expedition end this way. The grave of *PT 109* is marked by a torpedo headstone. Two young Americans died in the service of their country. This is their memorial.

ALTHOUGH MY SCHEDULE BEFORE I RETURN TO THE U.S. IS FULL, I'll be visiting Biuku and Eroni at Biuku's home again, and I'm going to climb up the slope of Kolombangara to the point where Reginald Evans watched the flames of *PT 109* from his coastwatcher station. The scientific part of this expedition is over. We set a very hard goal for ourselves, and at the 11th hour, using equipment that was not ideal, we accomplished our goal.

With the rain squalls rolling across the Blackett Strait and the clouds hanging low on the volcanic cone of Kolombangara, I declare this expedition a success. I feel connected to those momentous events of 59 years ago, when the world was at war and young men were called on to make a sacrifice that we hope our own sons and daughters will not often have to do.

I offer a salute to John F. Kennedy and his crew, and to those on both sides who came face to face with death and grew into manhood here in the mysterious Solomon Islands.

WE HAD TO AWAIT THE FINAL VERDICT of the Navy before we would tell the world what we'd accomplished. In mid-June, in steamy Washington, D.C., a panel of experts was duly convened to deliver their judgment. The panel consisted of Mark Wertheimer, Assistant Curator of Weapons, Naval Historical Center at the Navy Museum; Claire Peachey from the Center's Underwater Archaeology Branch; and Welford West, former torpedoman on *PT 157*, the boat that had rescued Kennedy and his men on August 8, 1943.

These three experts came to the National Geographic Society and watched our HDTV videotapes. They debated, they considered, and they reran the tapes. And though they were not at all convinced coming into the room that we had found *PT 109*, after looking at our visual evidence and at the documented absence of the sinking of any other similar vessel or equipment in that part of the Solomon Islands, they gradually came to the same conclusion I had: We had found the torpedo launcher and torpedo from an Elco-class PT boat, identical to *PT 109*.

Mark Wertheimer wrote to me a few days later with an interesting theory:

"We are still on the job researching why the torpedo and tube appear broken. Here is some speculation: As the boat sank, the torpedo shifted in the tube, and tripped the start latch, while inside the tube. However, the torpedo may have been still strapped in; the impulse charge would not have been fired. ... Therefore, you have what is known as a 'hot run,' with the torpedo running but able to go nowhere.

"As the torpedo energy builds up, the breech end of the tube (the back end) begins to heat up, and eventually the torpedo may break in some way (the one found is broken around the fuel flask section) and the tube melts away the tube or breaks open."

Mark Wertheimer wrote in his final report to the Navy:

"Conclusions: Physical evidence would strongly suggest that this is likely part of the *PT 109*. No serial numbers or other features could be discerned to make a definitive identification of the wreckage's affiliation with the boat. National Geographic and Navy research indicates no other PT boats were lost in this area.

"Recommendations / follow on action: Concur with National Geographic claims that this is likely *PT 109*.

"Statement: The discovery of this wreck is an important finding for the US Navy and the nation. Its likely affiliation with a future President will provide further physical evidence of his contribution to the prosecution of wartime missions."

Our find had become official. I would have liked to have the ship's nameplate in floodlights and JFK's ship log and that 37-mm gun that had never had a chance to fire at the Japanese. What I had was a barnacle-encrusted torpedo tube, but it was the real thing.

EPILOGUE

JOHN F. KENNEDY PROVED HIMSELF strong and brave and clever and resourceful. War is an unkind crucible of discovery; there is usually no middle ground. Seeing two of his men dead and gone, and others burned or injured, had spurred him to go beyond the call of duty. Swimming mile upon mile while exhausted, exercising caution when being more reckless might have led to Japanese discovery, he brought his men back . By the time of his rescue August 8, John Kennedy was a hero and a man in his own right.

IT'S MY INTENTION TO HAVE *PT 109* BE THE LAST of my contemporary shipwreck programs. When all is said and done, finding something that we already knew about is not the same as discovering something we know little about. My work in deepwater archaeology can have much greater scientific significance.

In that spirit, I'll be returning to the academic world. My new appointment at the University of Rhode Island's Graduate School of Oceanography will be in addition to my continuing work at the Institute for Exploration in Mystic.

I can see a number of goals before me right now. They include the establishment of a Ph.D. program in deepwater archaeology, completing my development of our family of remotely operated vehicles, and using tele-presence technology.

I want to turn my exploration sights ever deeper into our past. To the Black Sea, and maybe the flood that led to the story of Noah. To Inca treasure at the bottom of a lake in the Andes. To the lost exploded island of Santorini. Back to the very origin of who we are as a people. Back to the time when myths were born, and perhaps to the physical evidence to separate myth from history.

And so, in my own way, I change course. A little mid-course correction, you might say, because I see lots of adventures ahead. I believe *Titanic* and *Yorktown* and *PT 109* are just the beginning.

Thanks for coming with me this far on the journey. I promise you an even wilder and more wondrous ride to come.

Robert D. Ballard,
Lyme, Connecticut, August 2002

DECLASSIFIED "REPORT ON LOSS OF PT-109"

CMTB/L11-1
Serial 006
Declassified (8 SEP 59)
From: Commander, Motor Torpedo Boat Squadrons, South Pacific Force.
To: Commander-in-Chief, United States Fleet.
Via: Commander, South Pacific Force.
Subject: Loss of PT-109—Information concerning.
Reference: (a) ComSoPac's secret ltr. L11-1(11) Ser. 002867 of 30 December 1943.
Enclosure:
(A) Copy of ComMTB Rendova action report of 1-2 August 1943.
(B) Copy of ComMTB Rendova action report of 7-8 August 1943.
(C) Copy of Intelligence Officers' Memo to ComMTB Flot One of 22 August 1943. [NOT INCLUDED HERE]
1. Enclosures (A), (B), and (C) are forwarded in compliance with directive contained in reference (a). 2. Enclosures (A) and (B) are copies of action reports of Commander, Motor Torpedo Boats, Rendova, and contain information in connection with the loss of the PT 109. Enclosure (C) is a memorandum compiled by Intelligence Officers of Motor Torpedo Boat Flotilla ONE on the basis of information given them by survivors of PT 109. It is the most detailed account of this incident and it is hoped that it will provide the information requested in Enclosure (A) to reference (a).
E. J. MORAN.
W. C. SIECHT,
By direction.
Enclosure (A)

MOTOR TORPEDO BOATS, RENDOVA
5 August 1943.
MTBR/A16-3
Serial 0034
Declassified (8 SEP 59)
From: The Commander.
To: The Commander-in-Chief, U.S. Fleet.
Via: Official Channels.
Subject: PT Operations night 1-2 August 1943.
1. Force: All available boats (15) on patrol.
2. Enemy contracts: Five enemy destroyers, attacked in Blackett Strait, five or possibly six torpedo hits scored.
3. Weather: Overcast, visibility poor.
4. Patrols:

AREA B (BLACKETT STRAIT)
DIVISION B - OFF VANGA VANGA

Lt. H. J. Brantingham	PT 159	OAK 27
Lt. (jg) W. F. Liebenow	PT 157	OAK 21
Lt. (jg) J. R. Lowrey	PT 162	OAK 36
Lt. (jg) Jack Kennedy	PT 109	OAK 14

DIVISION A - OFF GATERE

Lt. A. H. Berndtson	PT 171	OAK 44
Lt. (jg) P. A. Potter	PT 169	OAK 31
Lt. (jg) S. Hamilton	PT 172	OAK 47
Ens. E. H. Kruse	PT 163	OAK 19

DIVISION R - EAST OF MAKUTI ISLAND

Lt. R. W. Rome	PT 174	OAK 50
Lt. (jg) R. E. Keresey	PT 105	OAK 7
Lt. (jg) R. K. Roberts	PT 103	OAK 1

DIVISION C - SOUTH OF FERGUSON PASSAGE

Lt. G. C. Cookman	PT 107	OAK 13
Lt. (jg) R. D. Shearer	PT 104	OAK 4
Lt. (jg) D. M. Payne	PT 106	OAK 10
Lt. (jg) S. D. Hix	PT 108	OAK 16

INCOMING TOKYO EXPRESS
All boats on the stations above indicated by 2130.
At 2400 Division B made radar contact indicating 5 craft approaching from the North close to the coast of Kolombangara Island. Visual contact was made shortly thereafter, by PT 159 which saw 4 shapes in column heading Southeast close into the coast at 15 knots. The PT 157 saw only two. The shapes were first believed to be large landing craft. The PTs 159 and 157, after directing the PTs 162 and 109 to lay to, began closing to make a strafing attack. In a moment the enemy opened fire with many large caliber guns, which was continued for several minutes. PT 159 fired a spread of 4 torpedoes and the PT 157, 2 torpedoes, all at a range of about 1800 yards. The torpedo tubes of the PT 159 flashed and one caught fire. A large explosion was seen at the target by personnel on both of these boats. They then retired to the Northwest laying puffs of smoke and making frequent radical course changes, until they were in Gizo Strait, where

they lay to. It was decided that PT 157 should return to station and that the PT 159 should return to base, as it was out of torpedoes, all of which was done. PTs 162 and 109 lay to as directed. When the firing began, there was so much and over such a long stretch of coast, they thought shore batteries had opened up and retired to the Northwest, but did not regain contact with the other two boats. After the firing had ceased, they were joined by PT 169 from Division A, and after receiving radio orders to do so, took up station, but did not make contract with PT 157. The PT 169 stayed with the PTs 162 and 109 on Division A's station off Vanga Vanga.
DIVISION A: Around 0004 Division A picked up 4 destroyers headed close in shore off Gatere. When PT 171 got in position it was abeam the first destroyer. Estimating its speed at 30 knots, the PT 171 closed to 1500 yards, at which point the destroyers fired starshells and opened fire, straddling the PT 171 and splashing water on its deck. Fire was also opened with automatic weapons and one destroyer turned on its searchlight but did not pick up PT 171. The PT 171 let go 4 torpedoes at the second destroyer. The tubes flashed and the destroyers turned directly toward it to evade. One destroyer stood on South toward Ferguson Passage. The last destroyer was soon to drop 2 1/2 miles behind the others. The PT 171 retired to the South laying smoke puffs and then getting out from behind them to the right and left. Feeling that the first destroyer might be blocking Ferguson Passage the PT 171 reversed course and proceeded Northwest along the reefs to the East of Gizo and out Gizo Passage departing for base, having expended all its torpedoes. The other three boats, PTs 170, 169 and 172 did not receive the contact report or any message to deploy for attack and could not fire their torpedoes after the destroyers opened fire, as PT 171 was in the way crossing their bows in its turn to the South. Contact between PT 169 and the other 3 PTs was lost as it reversed course to the Northwest after hearing radio message that destroyers might be blocking Ferguson Passage. After proceeding some distance North, (where it joined the PTs 159 and 157), the PTs 170 and 172 were straddled by the gunfire from the 2 destroyers, which they saw, but could not fire at because PT 171 was in front of them, retired zig-zagging and laying smoke puffs to the South thru Ferguson Passage. Going thru they were attacked by 4 float planes which dropped 3 flares and 2 bombs, which missed. They proceeded to the South and East, but returned to station on orders at 0255. Nothing further happened.
DIVISION C: When enroute to station Southeast of Gizo Island this section was circled by planes. At 0005, 2 ships were picked up by the radar on the PT 107. No previous contact report had been received, but a searchlight and gunfire had been seen to the North. PT 107 proceeded at high speed thru Ferguson Passage to attack, leaving the other 2 PTs behind. Inside of Ferguson Passage the PT 107 fired a spread of 4 torpedoes by radar. Shortly thereafter a dull red flash was seen in the direction of the target. Course was reversed and the PT 107, apparently undiscovered, proceeded South thru Ferguson Passage, enroute to base, its fish expended. PTs 104, 106, and 108 coming North thru the Passage were passed. In Ferguson Passage all these boats were attacked by a plane, which dropped flares and bombs and attempted to strafe the boats. There were no injuries or damage by this attack. PTs 104, 106, and 108 had no radar and saw nothing to fire at, however, they proceeded into Blackett Strait where they saw an explosion East of Makuti Island. They patrolled until ordered to resume their original station at 0137. The rest of their patrol was negative.
DIVISION R: At 0010 this division saw gun flashed to the North which continued for about 10 minutes, 3 or 4 much larger flashes were seen. A flare and bombs were dropped about a mile west of them a moment later. At 0025 the shape of ship was seen by PT 174 to the Northeast lying to or moving very slowly, about one mile off the shore of Kolombangara Island and seemed to be guarding the entrance to Blackett Strait from Ferguson Passage. She was firing at something to the West using a searchlight, all of which illuminated her. PT 174 fired 4 torpedoes at 1000 yards range and two explosions were seen at the target. The PT 174 circled to the right, passed behind Makuti Island and headed for Ferguson Passage. The ship fired shells which passed overhead, so the PT 174 used smoke puffs and put on speed. A plane also made a strafing run on it. PT 174 then proceeded to base as it had expended all its torpedoes. PT 103 sighted the destroyer when it turned on its searchlight. It had no previous information of contact with the enemy. After the PT 174 fired, the PT 103 also fired 4 torpedoes at a range of 2 miles. One flash was seen about 3 minutes later and possible a second. They retired at slow speed but when shells hit about 150' behind they increased speed and used smoke puffs. The passed out behind Makuti Island and headed for base, having fired all torpedoes.
PT 105 sighted the destroyer when it turned on searchlight and

began firing to the West, but had not received the previous advice that the enemy was in the area.

OUTGOING TOKYO EXPRESS, DIVISION R. As related PT 105 with 2 torpedoes was on duty at Ferguson Passage patrolling just inside on one engine. Just before 0230 a flame flashed up to the Northwest in the middle of Blackett Strait opposite Gatere. Gunfire immediately broke out about a mile to the North along the Kolombangara Coast. All of this showed the outline of a destroyer 2000 yards away to the East moving slowly to the North at about 10 knots. PT 105 was abeam this vessel. Two torpedoes were fired, but no explosion was observed. The PT 105 retired to the South and radioed that a target was proceeding North,. As the PT 105 passed thru

Ferguson Passage heading for base, its fish expended, three PTs were seen headed North thru the Passage.

DIVISION C. PT 107 had left for base, all its torpedoes gone, PTs 104, 106, and 108 had resumed station south of Ferguson Passage. When the explosion and firing were seen to the North around 0215, these three PTs went back through Ferguson Passage into Blackett Strait, but were unable to find anything.

DIVISION A. PTs 172 and 163 which had retired well to the South of Ferguson Passage did not resume station until after 0255 and were too late to make contact. PT 169 was with Division B to the North in Blackett Strait.

DIVISION B. As hereinbefore set out, PTs 162 and 109 of Division B with PT 169 of Division A were in Blackett Strait off Vanga Vanga, as was PT 157, which however, was not in contact with them. Around 0215 the three were due East of Gizo Island headed South, in right echelon formation with PT 109 leading, PT 162 second and PT 169 last. PT 162 saw on a collision course, a warship headed Northward about 700 yards away. The PT 162 turned to fire its torpedoes, but they did not fire. The PT 162 finally turned to the Southwest upon getting within 100 yards of the warship, to avoid collision. Personnel aboard the PT 162 saw 2 raked stacks, and at least 2 turrets aft, and possibly a third turret. At the time of turning, PT 109 was seen to collide with the warship, followed by an explosion and a large flame which died down a little, but continued to burn for 10 or 15 minutes. The warship when it was about 3000 yards away headed toward them at high speed. The PT 169 stopped just before the warship hit PT 109, turned toward it and fired two torpedoes when abeam at 150 yards range. The destroyer straddled the PT 169

with shell fire, just after it a collision with PT 109, and then circled left toward Gizo Island at increased speed and disappeared.

The PT 169 laid smoke screen and zigzaged to the Southeast along the reefs off Gizo Island. About 0245 a wake was seen coming up from the near Northwest and on a parallel course. The PT 169 swung around to the left toward the ship (a destroyer) and fired port and starboard forward torpedoes at 2000 yards. The destroyer turned to its port just in time for the starboard torpedo to hit its bow and explode. The PT 169 continued its swing and retired South thru Ferguson Passage going at high speed for 1/2 mile laying smoke and zigzaging and headed for base. All its torpedoes gone.

PT 157 was farther North than the other 3 PTs. About 0200 the PT 157 saw a ship close in shore off Kolombangara due East of the center of Gizo Island and fired 2 torpedoes at it, but no explosion was seen. The ship continued Northwest at about 5 knots, without firing and disappeared.

No further contact was made with the express. The boats remaining on station departed for base at 0400.

5. All times are Love.

6. COMMUNICATIONS: Communications with base were good, however, several PTs failed to put out immediate intelligible report of contact with the enemy, with the result that the others had no chance to get into position for an attack.

7. COMMENTS AND RECOMMENDATIONS:

(a) Contact reports giving the senders call, the type, position, course and speed of the enemy should be radioed immediately in plain language. PTs not making the contact should refrain from all radio traffic themselves (except contact reports) until all reasonable possibility of making contact has ended. The boats making contact should continue reports of enemy position, etc. after torpedo firing and as long as the enemy is visible or on radar.

(b) PTs should stay together in "V" formation and follow their division leader. All boats should fire their torpedoes when their section leader fires, without deployment. They should spread torpedoes about the base torpedo course of the leader.

(c) The boats should fire at shorter range. Some boats retired without firing and had to be directed to return to station.

(d) The Mark VIII torpedo again manifested its want of capacity to

inflict real damage. Enemy destroyers kept going after certain hits had been scored. Intelligence reports that 5 unexploded torpedoes are on the shore of Kolombangara Island.

(e) Flashes and burning in the tubes on firing not only give target opportunity to avoid but disclose PT positions. Not enough interest is being taken in this matter behind the firing line.

T. G. WARFIELD.

Advance copy to:
CicPac
CTF 31
Copy to:
CoMTB, Russells
CNB, Dowser
ComAirSols
ComAir, New Georgia
ComGen, New Georgia
Enclosure (B)

MOTOR TORPEDO BOATS, RENDOVA
8 August 1943.
MTBR/A16-3
Serial 0038 (1)
Declassified (8 SEP 59)
From: The Commander.
To: The Commander in Chief, U.S. Fleet.
Via: Official Channels.
Subject: PT Operations night 7-8 August 1943,

1. FORCE: Eight boats on patrol. Two boats on the alert at base.

2. ENEMY CONTACTS: None.

3. WEATHER: Overcast with occasional showers, visibility poor to fair.

4. PATROL:

AREA K (EASTERN SIDE OF LOWER VELLA GULF)
DIVISION A

Lt. A. H. Berndtson	PT 171	OAK 44
Ens. W.F. Criffin	PT 168	OAK 28
DIVISION P		
Lt.(jg) D. M. Payne	PT 106	OAK 10
Lt.(jg) R. D. Shearer	PT 104	OAK 4
DIVISION T		
Lt.(jg) P. A. Potter	PT 169	OAK 31
Lt.(jg) J. E. McElroy	PT 161	OAK 33
DIVISION D		
Lt.(jg) D. S. Kennedy	PT 118	OAK 6
Lt.(jg) H. D. Smith	PT 154	OAK 12

All boats arrived on station 2130, except PT 171 which waited to accompany PT 157 on the rescue mission hereinafter set out. No contacts or sightings were made during the night by radar or otherwise. Left for base at 0330. Enroute to base a raft with two Japs on it, made of two empty oil drums and planks, was found five miles West of Rendova Harbor. The Japs were taken prisoner and after delivered with all their gear to Army Intelligence, Commanding General, New Georgia.

5. COMMUNICATIONS: Communications were good with base, but transmissions between some of the boats were weak.

6. RESCUE MISSION: August 6, word was received from the Coastwatcher and by Native Messenger that eleven survivors of PT 109, sunk in a collision with an enemy destroyer on the morning of August 2, were alive and on a small Islet near Cross Island on the West side of Ferguson Passage. Arrangements were made through the Offices of the Coastwatcher Organization for the rescue. Lt.(jg) W. F. Liebenow in PT 157, assisted by PT 171 of the regular patrol made the rescue. The Native messengers were taken along as guides and were most helpful in guiding the PT 157 through the reefs and in handling the small boats. Several trips from shore over the reef to the PT 157 were required to remove all the men, three of whom had been badly, but not critically burned. Everything went off smoothly. The natives had fed and done everything to make the men comfortable during their stay on the island. PT 157 returned to base at 0500.

7. All times are Love.

T. G. WARFIELD.

Advance copy to:
CincPac
CTF 31
Copy to:
CoMTB, Russells
CNB Dowser
ComAirSols
ComAir, New Georgia
ComGen, New Georgia

INDEX

190

191

PHOTO CREDITS

Key: JFKL = John F. Kennedy Library, Boston

Front Matter: p. 2, Tom Freeman; pp. 4-5, JFKL, #PC100; p. 9, JFKL, #PC164.

Chapter One: p. 12, JFKL, #PC92; p. 15, Robert Ballard/Odyssey Enterprises; pp. 16-17, Henry Godines, Avant-Garde Publishing; p. 21, Ira Block.

Chapter Two: p. 26, JFKL, #PX81-32:152; p. 29 (upper), JFKL, #KFC618N; p. 29 (lower), JFKL, #PX81-32:329; pp. 32-33, Dorothy Wildling, by courtesy of the National Portrait Gallery, London; pp. 35 & 36, AP / Wide World Photos; pp. 38-39, JFKL, #PX65-137; p. 41, JFKL, #KFC1950P; p. 42, JFKL, #PC95; p. 45, JFKL, #PX77-3; pp. 46-47, JFKL, #PX73-52:6; p. 49, JFKL, #PX73-52:13; pp. 50-51, JFKL, #PX73-52:5; pp. 55 & 56, Hulton|Archive by Getty Images; p. 59, U.S. Marine Corps, Official Photograph; p. 60, Hulton|Archive by Getty Images; pp. 62-63, U.S. Coast Guard, Official Photograph; pp. 66 & 67, Naval Historical Foundation; pp. 71 & 72, CORBIS; p. 77, Hulton-Deutsch Collection/CORBIS.

Chapter Three: p. 80, JFKL, #PC101; pp. 84-85, Tom Freeman; p. 89, JFKL, #POF/PSF/PT109:2; pp. 90-91, JFKL, #POF/PSF/PT109:4; pp. 94-95, Tom Freeman; p. 96 (upper), Elliott Erwitt/Magnum Photos; p. 96 (lower), U.S. Naval Historical Center; p. 99, Elliott Erwitt/Magnum Photos; pp. 102-103, Ira Block; p. 104, Elliott Erwitt/Magnum Photos; p. 108 (upper), JFKL, #PX73-50:1; p. 108 (lower), JFKL, #FY-AY1; p. 109, Ira Block; pp. 112-113, JFKL, #PC2239; p. 116 (upper), JFKL, #PC102; p. 116 (lower), JFKL, #PC96; p. 121, JFKL, #PC109; p. 122, AP/Wide World Photos; p. 123 (upper), JFKL, #PC173; p. 123 (lower), JFKL, #PC165; p. 124, JFKL; p. 127, Elliott Erwitt/Magnum Photos.

Chapter Four: p. 130, Ira Block; pp. 133 (both) & 134, H. Ian Hogbin, courtesy Australian Academy of Science; pp. 135 & 138-139, Ira Block; p. 140, H. Ian Hogbin, courtesy Australian Academy of Science; p. 143, Hulton|Archive by Getty Images; pp. 146-147, Ira Block; p. 148, Pascal Maitre/Cannelle Agency; p. 151, Mike McCoy.

Chapter Five: pp. 154-171 (all), Ira Block; pp. 174-177 (all), Odyssey Enterprises/ Institute for Exploration; pp. 181 & 184-185, Ira Block.

**PUBLISHED BY THE
NATIONAL GEOGRAPHIC SOCIETY**

John M. Fahey, Jr.
President and Chief Executive Officer

Gilbert M. Grosvenor
Chairman of the Board

Nina D. Hoffman
Executive Vice President

PREPARED BY THE BOOK DIVISION

Kevin Mulroy, *Vice President and Editor-in-Chief*

Charles Kogod, *Illustrations Director*

Marianne R. Koszorus, *Design Director*

STAFF FOR THIS BOOK

Johnna M. Rizzo, *Editor*

Sadie Quarrier, *Illustrations Editor*

Gerry Greaney, *Art Director*

Carl Mehler, *Director of Maps*

Matt Chwastyk, Joe Ochlak, Nick Rosenbach,
Greg Ugiansky, and National Geographic Maps,
Map Research and Production

Alexander Feldman, *Researcher*

Gary Colbert, *Production Director*

Richard S. Wain, *Production Project Manager*

Meredith C. Wilcox, *Illustrations Assistant*

Connie Binder, *Indexer*

**MANUFACTURING AND QUALITY
CONTROL**

Christopher A. Liedel, *Chief Financial Officer*

Phillip L. Schlosser, *Managing Director*

John T. Dunn, *Technical Director*

Vincent P. Ryan, *Manager*

One of the world's largest nonprofit scientific and educational organizations, the NATIONAL GEOGRAPHIC SOCIETY was founded in 1888 "for the increase and diffusion of geographic knowledge." Fulfilling this mission, the Society educates and inspires millions every day through its magazines, books, television programs, videos, maps and atlases, research grants, the National Geographic Bee, teacher workshops, and innovative classroom materials. The Society is supported through membership dues, charitable gifts, and income from the sale of its educational products. This support is vital to National Geographic's mission to increase global understanding and promote conservation of our planet through exploration, research, and education.

For more information, please call 1-800-NGS LINE (647-5463) or write to the following address:
NATIONAL GEOGRAPHIC SOCIETY
1145 17th Street N.W, .Washington, D.C. 20036-4688, U.S.A.

Visit the Society's Web site at www.nationalgeo-graphic.com.

Printed and bound in U.S.A. by
R. R. Donnelley & Sons, Willard, Ohio.
Separations by Quad Graphics,
Alexandria, Virginia.
Dust jacket printed by Miken Companies,
Cheektowaga, New York.

Copyright © 2002 Odyssey Enterprises, Inc.
First printing, November 2002

Library of Congress Cataloging-in-Publication Data

Ballard, Robert D.
  Collision with history : the search for John F. Kennedy's PT 109 / Robert D. Ballard,
with Michael Hamilton Morgan.
    p. cm.
  Includes index.
  ISBN 0-7922-6876-8
  1. PT-109 (Torpedo boat) 2. World War, 1939-1945--Campaigns--Solomon Islands. 3.
Kennedy, John F. (John Fitzgerald), 1917-1963--Military leadership. 4. World War,
1939-1945--Naval operations, American. 5. Shipwrecks--Solomon Islands. 6.
Underwater archaeology--Solomon Islands. 7. Excavations (Archaeology)--Solomon
Islands. 8. Solomon Islands--Antiquities. 9. Ballard, Robert D. I. Morgan, Michael
Hamilton. II. Title.

D774.P8 B35 2002
940.54'5973--dc21

               2002032141